THE
LIFE-CHANGING
MAGIC OF
NUMBERS

BOBBY
SEAGULL

1 3 5 7 9 10 8 6 4 2

Virgin Books, an imprint of Ebury Publishing,
20 Vauxhall Bridge Road,
London SW1V 2SA

Virgin Books is part of the Penguin Random House group of companies
whose addresses can be found at global.penguinrandomhouse.com

Penguin
Random House
UK

First published in the United Kingdom by Virgin Books in 2018
This edition published in the United Kingdom by Virgin Books in 2019

www.penguin.co.uk

A CIP catalogue record for this book is available from the British Library

ISBN 9780753552803

Printed and bound in Great Britain by Clays Ltd, Elcograf S.p.A.

Penguin Random House is committed to a sustainable future for our
business, our readers and our planet. This book is made from Forest
Stewardship Council® certified paper.

MIX
Paper from
responsible sources
FSC
www.fsc.org   FSC® C018179

*To my family without whom I would not be
the person I am today.*

Contents

CHAPTER 1

'It's Easy as 1, 2, 3'

Counting, Ordering Numbers and Basic Data Handling Tools

It was the winter of 1993, Meat Loaf's 'I'd Do Anything For Love' was number 1 in the charts, and I was a small boy of 9 standing in the playground in an oversized hooded winter jacket, clutching my most prized possession: the Merlin 1993–94 Premier League football collection sticker book. Growing up as a child in 1990s east London, the language of the playground was football and the currency to participate was football stickers.

'Got, got, got, need.' This was a phrase that frequently passed my lips, as I attempted to barter stickers with other children. I calmly stated the stickers that I had already 'got' in my collection, but excitedly blurted out 'need' for the stickers that were still required. After the school bell rang out for the weekend, my regular Friday afternoon rush was to the local corner shop. I would spend my few pounds of pocket money on several packets of overpriced adhesive labels.

My friends not only traded stickers, but we used to hypothesise about which players would make the ideal transfer. While Blackburn Rovers' prolific striker Alan Shearer would be more suited to Alex Ferguson's forward line at Manchester United, children could dream of a world where Alan would ply his trade at West Ham's Upton Park. The sticker world was clearly not the real world!

This is where numbers first played a role in my social identity. Of course, my first connection with numbers was that most simple pleasure, one of counting basic numbers: 1, 2, 3 ... However, it was in this new setting that I found numbers coming into their own. Most footballing conversations with my peers were based on assertions without evidence – such as saying that Arsenal's Ian Wright was a more effective striker than Southampton's Matt Le Tissier – but the confidence to use numbers helped me to carve my own identity. I knew that these strikers had both scored 15 goals in their previous league campaigns and this fact helped to add credence to my arguments.

After swapping new stickers, I recall rushing home with my sticker book and furiously tapping in the information on all my new stickers into my master spreadsheet in Microsoft Excel. For example, with Manchester United's striker Eric Cantona, I entered his personal details (height 1.86 metres, weight 81 kilograms) and his performance statistics (33 appearances; 2 substitute; 15 goals, of which 10 were with the right foot, 1 with the left foot and 4 with the head). A few years later in secondary school, I progressed onto using the database software Microsoft Access for this same task but with more efficiency. Inputs included players' names, starting position, heights and data from the previous year, such as games played and goals scored.

Armed with this, I was able make more accurate assessments about players. If a West Ham fan wanted to find out possible players who were defenders, more than 1.83 metres and played more than 20 games in the season prior, a quick sorting formula on Excel would weave its magic. And bam! I would be able to come up with a list of a few players who could fit the bill.

I found the relatively painless skill to rank players by different criteria in my spreadsheet – height, goals scored or games played – allowed me to turn a complex world into something I could begin to fathom. This balanced my subjective opinions of players based on their trickery and confidence on the ball with objective comparisons of players. Apparently meaningless numbers, such as the heights of various players, could be brought to life with basic data techniques such as mean, median and modes that let me compare them against each other. Suddenly I could impress my friends with answers to questions like 'Which is the average tallest squad in the league?' or 'Which team has the highest number of goals being contributed from midfield?' All straightforward to interrogate within a few clicks of Excel. Perhaps I missed a trick in not turning this objective analysis of football into a business – now companies like Opta Sports have done so with shrewd commercial success!

What numbers gave me was a sense of 'knowing' in a world of conjecture. Like Linus's security blanket in the Snoopy cartoons, numbers gave me a bedrock upon which to form myself. In fact, if I'm really honest my sense of the certainty of data can be traced back a couple of years earlier than the Premier League sticker book. It stemmed from a Christmas gift in 1991 and a book that perhaps changed my life and rooted my love of numbers, data and information: *The Top 10 of Everything 1992* by Russell Ash. From the moment I delicately unwrapped the paper from this gift, I was devouring ordered lists. Lists about the Top 10 tallest buildings in the world, the Top 10 grossing films at the box office, the Top 10 longest migration paths of any species, and more. You name a topic – astronomy, art, food and drink, music – and I could start reciting off lists at breakneck speed (perhaps without quite understanding what lay behind the numbers).

The 1992 Barcelona Olympic summer games was the first time I saw a visually captivating way numbers could express the physical peak of young athletes: Team GB captain Linford Christie storming through to win the 100 metres in 9.96 seconds; Carl Lewis edging out Mike Powell by 3 centimetres in an epic men's long jump final; Lavinia Miloşovici's mesmerising floor routine in the gymnastics to secure her a perfect 10.0 score. As a little boy, I was sprawled on the living room floor, scoffing my face with digestive biscuits and fixated by the sports. I was slowly coming to understand that numbers were a way of working out sizes and sequences, and establishing who the winners and losers were. Numbers were abundant in my eyes, ranking and ordering competitors and nations, trying to objectively quantify things that seemed a million miles from my maths lessons at school.

Early on, I had learned to adore numbers not just in the classroom, but also in the real world. 'But what's the big deal with numbers?' you might ask. 'Where have they come from? And what's the point in ordering them and processing them into neat little sets of information?'

In the beginning God created the Heaven and the Earth, and if you ask me, He created numbers shortly after, especially if He needed to take the seventh day off as rest. There is evidence to suggest that within a few hours of being born, babies start developing an abstract representation of numbers.

It is at this abstract stage where we as humans move beyond most other species on the planet. Many animals do seem to distinguish, in a primitive manner, between 'more' and 'less' as a concept. But as a species, humans have elevated ourselves beyond hunter-gatherers, for the most part, to a more enlightened existence (scrolling through cute cat videos on Facebook). However, deep in the Amazon rainforest, the Pirahã tribe continue to live as they seem to have for time immemorial. Their counting system goes '1', '2' and then 'many'. In a similar vein, the African Hadza tribe, living near the central Rift Valley in Tanzania, count only up to 3 and no more. Both these tribes seem to operate their society without the need to use specific large numbers, such as 107. However, the vast majority of humanity continues to progress by establishing relationships between abstract concepts of numbers involving the simple arithmetic operations of adding, subtracting, multiplying and dividing.

Humans, assuming no mutation, are born with 10 digits (10 fingers and thumbs in total), providing us with a natural counting frame. Given this, it's no coincidence that many civilisations developed a decimal (base 10) counting system. The Latin word for finger (*digitus*) means a number. Students I teach have asked me what would happen if humanity had been born with 3 fingers per hand, a bit like the Teenage Mutant Ninja Turtles. With a total of 6 fingers, then perhaps we would have developed a counting system with base 6 – and perhaps a natural affinity for pizza and throwing majestic martial arts moves!

Let me explain how base counting works. Our base 10 decimal counting system is the one we're all familiar with and, based on groupings of 10, it uses different symbols (0, 1, 2, 3, 4, 5, 6, 7, 8, 9) to represent all the whole numbers we know. We then group numbers in batches of 10, as they are divisible by this base of 10 – such as 10, 20, 30, 100, 500, 100,000 – and they are reflected in our collective nouns such as decade, century, millennium. It's intriguing to think that collective nouns we take for granted, such as a century in cricket or my *Top 10 of Everything* book, are an arbitrary choice, based on our decimal system. It would be equally as valid for the Teenage Mutant Ninja Turtles to publish their *Top 6 of Everything* book (where in this alternate universe, a young Roberto the turtle learned to admire maths and enjoyed compiling lists of 6!).

If we used base 6, our counting would follow this order: 0, 1, 2, 3, 4, 5, then pick up at 10 (which would be 6), 11, 12, 13, 14, 15, then pick up again at 20 (but this 20 would represent 12 items in our decimal system). So the next time you make a trade to purchase an item with someone and say you'll give them a tenner for that signed West Ham match-day programme, try this instead. Turn up with £6 and explain that you hadn't agreed the base of the currency and then be ready to run!

A classic pub quiz question is to identify which country has the highest number of living languages (not dialects, otherwise London itself might win!). Papua New Guinea, occupying the eastern part of the planet's second largest island, north of Australia, holds this record with around 841 according to the 'Ethnologue: Languages of the World' study in 2018. Not coincidentally perhaps, the tribes of this country also stake the claim for the most number of counting systems, perhaps 900. The Alamblak only have words for 1, 2, 5 and 20 and all their other numbers are constructed from those. Another, the Oksapmin, use base 27 as they count on 27 different body parts. To describe numbers larger than 27, they add the increment of '1 man'. So to describe the number 50 would be 1 man (27) and the left thumb (which is 23).

In early human history, when people desired to keep a record of something (perhaps grain or other food stuff), notches in sticks or stones were the natural solution. Numbers were built up with a repeated sign for each group of 10, continued by repeated signs for 1. The ancient Egyptians represented 1 with a vertical line and 10 as a high-hat symbol (^). As they wrote from right to left, the number 25 would have been IIIII^^. In contrast, the Babylonians used 60 as the base of their numerical system, which required them to use a different sign for every single number up to 59 (in the same manner in which we need a different number for every digit up to 9 in our decimal system). The influence of the Babylonians' base of 60 still resonates today, particularly in our time systems (60 seconds and minutes in angular measurement) and geometry (180 degrees in a triangle and 360 degrees in a circle).

The most practical concept that the Babylonians advanced was that of 'place value'. This is where the same digit has a different value depending on where it occurs in the sequence. For example, for 444, the digit 4 means 400, 40 and 4 depending where it's placed. While this seems mundane to us now, this was revolutionary in Babylon.

The next leap forward was the invention of the zero as its own number, which seems to have been achieved in my ancestral homeland of India around the third century BC. Nowadays, the binary system with 0s and 1s (base 2) allows computers to communicate with each other at phenomenal speeds.

Counting is such a natural process to us that its copiousness is in plain sight all around us, even in our popular culture of music. From children we are exposed to nursery rhymes, from the classic '1, 2, 3, 4, 5, once I caught a fish alive' to '1, 2, buckle my shoe'. As children move onto adult music, numbers are still prevalent in lyrics – such as the 2017 viral music hit 'Man's Not Hot' by comedian-turned-rapper Big Shaq which opens '2 + 2 is 4, minus 1 that's 3, quick maths'.

Once we have counting numbers established, we are then able to proceed with making comparisons. If I have 2 toy cars and my cousin has 4 toy cars, then I can establish that I would need a further 2 cars to have the same number as him. With this, we can start ranking and ordering items per individual (e.g. in order of who has the most cars).

Various kinds of statistical techniques allow us to find meaning behind the data. To start with we learn basic numbers, initially the natural numbers (counting numbers such as 1, 2, 3, etc.). From there we start to understand negative numbers and 0 (-3, -2, -1, 0, 1, 2, 3) and expand our scope to integers. And beyond that we could move into the realm of rational numbers (numbers that can be expressed as fractions), and then irrational numbers, surds, prime numbers ... But these are for another journey.

But I can't resist talking about prime numbers right now. These are the building blocks of mathematics, much like atoms for chemistry or a palette of colours for artists. These are the numbers 2, 3, 5, 7, 11, 13 and so on. A prime number is a number greater than 1 that can only be divided by 1 and itself. So 17 is a prime number as it can only be divided by 1 and 17, whereas 10 is not a prime number as it can be divided by 2 and 5, as well as 1 and 10 of course.

The mathematical obsession behind primes partly lies in the fact that all non-prime numbers above 1 can be expressed as the multiplication of primes. For instance, 30 can be written as $2 \times 3 \times 5$ because 2, 3 and 5 are all primes. Likewise 123,456,789 is $3 \times 3 \times 3,607 \times 3,803$.

Beyond mathematical beauty, prime numbers underpin security in modern technology. The security of many encryption algorithms is based on the fact that while it is straightforward to multiply 2 large prime numbers and get an end product, it is incredibly challenging (even with a computer) to reverse this process. When we have a number that is the product of 2 primes (such as $21 = 3 \times 7$), it is easy as pie. But finding the 2 (or more) primes that multiply together for very large numbers is problematic. Indeed, the search for an algorithm that can perform this rapidly is one of the largest unsolved problems in computer science.

It has been impossible thus far for mathematicians to derive a truly efficient formula for factoring large numbers into primes. With small numbers such as 24, we can prime factorise this into $2 \times 2 \times 2 \times 3$ quite easily. We do have the technology to factor large numbers into primes, but if we start trying to use the same methods for 400- or 500-digit numbers, then even the most advanced computers in the world would take a ridiculously long time to solve. And we are talking about timescales that are longer than the age of Earth, and for ludicrously large numbers, perhaps even longer than the universe has existed.

So the next time you send a secure WhatsApp message, make an online transaction or even use a cash point, it is likely that large prime number multiplications are protecting you.

Even in the natural world, prime numbers are influential. There are cicada insect species on the east of the United States that leave their burrows in intervals of 7, 13 or 17 years. They have developed these natural rhythms through evolution, using prime number years of emerging to minimise overlapping with predators. In his bestseller *Ever Since Darwin* in 1977, the entomologist Stephen Jay Gould explained how potential predators often have 2–5-year life-cycles.

If a predator had a 5-year life-cycle, and cicadas emerged every 15 years, each cicada emergence would be devastated by the predators. However, by emerging at larger prime number cycles, these cicadas minimise their overlaps with these greedy predators. For example, with 17-year cicada cycles, they would only coincide with the predators every 85 years (17×5). As 13 and 17 are prime numbers, these year-length cycles cannot be tracked by any smaller number. It is truly remarkable that these insects use prime numbers to increase their chances of survival.

Recently I was at a session for Doctorate of Education candidates at Cambridge University to discuss the progress on their research on mathematics education. My personal research is on understanding the causes of maths anxiety. One of the other doctoral candidates, who was very close to being able to submit his final thesis, shared his experiences in his capacity as a former head of a maths department recruiting prospective maths teachers to his secondary school. He asked me 3 questions about data handling in quick-fire succession that he often asked candidates:

First question, 'What is the mode?'
My reply: 'The mode is the value that appears the most in a list of numbers.'
Second question, 'What is the median?'
'The median is the middle value when a list of numbers is placed in ascending order (from smallest to largest).'
And the final question, 'What is the mean?'
I said, 'The mean is the sum of all the numbers in a list divided by how many numbers there are.'

I had fallen into a classic trap. Yes, the mean can be calculated by the method I suggested but he had asked me for the *meaning* (no pun intended) of mean. Part of mathematics is about being precise, and I had answered by giving the procedure for the calculation of the mean, rather than the actual definition.

The mean is the average of a set of numbers and can be calculated to be the central value of the set of numbers. It is used to represent the data through a single value. The nuance is quite subtle, but when we calculate the arithmetic mean as discussed, this is a single value that can be used to represent the data. The *Oxford English Dictionary* defines the average as a 'number expressing the central or typical value in a set of data, in particular the mode, the median or (most commonly) the mean' (which can be calculated as previously stated).

So when we are asked about the average salary in a company, or the average life expectancy in a population, we are trying to find a number to express the typical value in that data set (the total number of employees in a company or the total size of a country's population). We usually refer to the arithmetic mean but as vigilant users of statistics, we should always be cognisant of which average we are using.

The phrase 'lies, damned lies and statistics' was popularised across the Atlantic by American author Mark Twain (among others). He attributed this to nineteenth-century British Prime Minister Benjamin Disraeli, who apparently said, 'There are three kinds of lies: lies, damned lies, and statistics.' So while the origin of this quote is disputed, the message it sends is that we need to be prudent when approaching numbers in an applied situation.

Imagine a scenario where I told you about 2 prospective jobs available. Two companies, A and B, each have 10 people who earn an average salary of £22k per year. If you received a job offer from both, you should in theory be indifferent to both. But this is where understanding how data and averages are used is crucial. First, the type of average used is the arithmetic mean: adding up the total salaries and dividing by the number of staff. However, upon further digging, you uncover that the reality is deceptive:

Company A has 9 staff on £10k per year and the boss on £120k per year (£10k, £10k, £10k, £10k, £10k, £10k, £10k, £10k, £10k, £120k).

Company B has 9 staff on £15k per year and the boss on £80k per year (£15k, £15k, £15k, £15k, £15k, £15k, £15k, £15k, £15k, £80k).

Now a clearer picture emerges. Company B is patently better for the majority of working staff, but the mean gives a misleading picture. Here, the mode or, even better, the median salary should enable you to make a better choice. Company B has a median of £15k and obviously is the more beneficial option for you (unless of course you are being offered the CEO's job!).

This micro picture with this example becomes crystal clear when looking at a macro picture of the UK economy. About two-thirds of the UK working population back in 2007 earned less than the *mean* salary. Think about that for a second. For that to be true, there must be some people among the third who earn more than the mean salary – dramatically more. Income distribution results in the very high earners distorting the mean and thus it's not a useful metric for this kind of data.

Instead, the *median* salary (the person in the middle) is a better indicator of what the 'average' person earns. As in my example, the outliers can damage the reliability of interpreting the arithmetic mean as a proxy for the whole population.

So our journey with numbers starts as we learn to count the natural numbers: 1, 2, 3 and so on. From there, we encounter negative numbers and how numbers can represent relationships between things even if we can't see them (we can see 2 toy cars, but not −2 toy cars). From here, we learn to make comparisons between sets of data. Trying to work out who's the quicker sprinter, which is the higher ranked football team or the better value house to buy. Numbers give us this objective framework by which to hang our subjective views. W.E.B. Du Bois, an American sociologist and civil rights activist, put it best when he said, 'When you have mastered numbers, you will in fact no longer be reading numbers, any more than you read words when reading books. You will be reading meanings.'

Chapter 1 Puzzle

Gareth Southgate's Unusually Weighted Footballs

Test your knowledge of statistical averages with the England football manager and waistcoat wonder Gareth Southgate. He has set his England team a mathematical challenge.

Gareth has given unusually weighted footballs to 4 of his squad members: captain Harry Kane, goalkeeper Jordan Pickford, and *Fortnite* computer game dance maestros Jesse Lingard and Dele Alli. Gareth has asked the players to work out the weight of each of the 4 footballs and has given them 4 clues.

Clue number 1: The mean or average weight of these footballs is 80 grams.

Clue number 2: The 3 heavier footballs weigh 270 grams in total.

Clue number 3: The heaviest football, given to Harry Kane, is 3 times the weight of the lightest football, given to Jesse Lingard.

Clue number 4: The 2 middle footballs weigh exactly the same.

What is the weight of each of the footballs?

CHAPTER 2

'If at First You Don't Succeed, Try, Try Again'

The Law of Large Numbers

Lucky Bobby wins trolley dash through new store.

Bobby Seagull, who won a competition to rush around the 99p Stores shop in High Street North in 99 seconds, admitted he has now bought most of his festive fare. Bobby, 27, from East Ham was the lucky winner chosen from 40 entries. He managed to scoop up 37 items in 99 seconds.

These are the opening lines to an article in my local newspaper back in December 2011. It describes how I triumphed in a competition to open a new 99p Store on my local high street, and won the prize of a mad 'trolley dash' to fill a shopping trolley with as many items as possible within 99 seconds, in my very own 99p Store Supermarket Sweep!

To enter this competition, entrants had to submit a 4 line poem that would be read out by the winner as they cut the red ribbon for the store. So, this was not purely a lucky draw competition, but involved an element of skill. Not a high degree of skill as demonstrated here by the cringeworthy poem that I gleefully (I have no shame) read out on a crisp December morning.

'Bargain Hunter'
Recession, recession – but I still need to make a good impression.
Trying to find that elusive bargain – sometimes involves a lot of pain!
But once that magical offer is found – sheer happiness is where I am bound.
At 99p Store – bargains galore!

Since the age of 5, I have always enjoyed entering competitions: some that involved no skill whatsoever (just submitting your details on a postcard), others with a limited bit of research (finding out the answers to a few general knowledge questions) and some with a greater or lesser degree of skill (from writing a poem to sending in artwork, or sending a photo of yourself dressed up as your favourite book character). With my postcards, it must be noted I tried to make them as distinctive as possible from other entrants by cutting the front of cereal boxes in half. Our local postman must have been bemused to see Corn Flakes, Rice Krispies or Frosties (or whatever breakfast cereal was in vogue at home at the time) as postcards flying through the mail every week or 2.

My apparent lucky streak has resulted in a dizzying array of prizes: meeting Lord of the Dance Michael Flatley on the red carpet of Leicester Square for his film's premiere; a luxury holiday to Belgium; a year's supply of chocolate; a high-spec laptop computer; a year's worth of travel insurance; a sky-diving opportunity (which I didn't manage to use before it expired!); and even a child's doll's house. For every prize mentioned above, I can name another 20 others. Was this a fluke? Have the gods of fortune been shining down upon me favourably? Was I simply born lucky?

I think not. My friends often comment to me that they 'never win anything. Bobby, you're just so lucky with competitions.' The reality is I exploit what is known in mathematics – and in particular, in a part of maths called probability – as the law of large numbers. Put simply, the more you do something, the more likely you will achieve your desired outcome. Whenever I used to watch swans swim gracefully through the rivers near Windsor, like most observers I noticed how elegant and poised the swans seemed. However, I understood that beneath the surface of the river, the swans' feet were manically rotating, giving the surface illusion of poetry in motion. Similarly with my competitions, the outsider observes a stream of successes but without realising I have entered scores of prize draws (although this was a hobby and something I enjoyed, so it was not an arduous ordeal!).

An early realisation of the power of probability (and indeed the power of infinity) came as a 9-year-old during one episode of *The Simpsons* in 1993. Watching this American cartoon was a particular treat, as it was only on Sky TV and involved a visit to my cousin's house (a brisk 5-minute walk from our house) and was possibly combined with some Premier League football and some of my auntie's delicious south Indian snacks. In this episode, Mr Burns, the owner of the Springfield Power Plant, has a room with a thousand monkeys working furiously at a thousand typewriters. Mr Burns was hoping that they would write the 'greatest novel known to man'. He then picks up the writing and reads out their efforts: 'It was the best of times, it was the blurst of times?' Realising that his fiendish plans weren't working, he throws the script at his 'stupid monkey'. (Bonus point if you spotted that this was meant to echo the opening lines to Dickens's *A Tale of Two Cities*.)

I understood the concept underlying Mr Burns's scheme and did some research at my local East Ham library, a beautiful red-brick Edwardian trove of knowledge where I spent many hours of my youth. In the pre-internet era, I had to consult the librarian and then leaf through books until I unearthed the idea behind this joke: the infinite monkey theorem. One of the first recorded uses of this metaphor was in 1913 by French mathematician Émile Borel. This theorem states that a monkey striking keys at random on a typewriter for an infinite amount of time will at some stage type the complete works of Shakespeare or in fact any given text, even this book you're reading right now! (I did suggest to my editor to find the smartest monkeys from London

Zoo and leave them to write this book, but he declined my offer saying that an infinite amount of time might cause problems getting the book out on schedule!)

The infinite monkey metaphor represents an abstract contraption that produces an endless random sequence of characters. Unfortunately, the probability that monkeys would type out a complete work, such as Isaac Newton's seminal 1687 work *Principia Mathematica*, is so minute that the chance of it happening during a period of time hundreds of thousands of orders of magnitude longer than the life span of the universe (13.7 billion years) is incredibly low (though technically speaking not 0). But the key message of the theorem is this: with enough time (even if many times the length of the age of the universe) and repetitions, all outcomes can occur, even getting monkeys to crack the most intricate encryptions known to mathematics.

What does this theorem mean for us? I think it demonstrates that if it is unlikely something will happen (or in mathematical terms, if it has a probability near 0), we can keep repeating it to increase the probability of it occurring, even if the probability still remains small. Crucially, this means that if you keep trying some-thing for long enough, it gets more and more likely that you'll achieve it. And that's why persistence really is the key to making your own luck. The harder you try, the luckier you get.

Let's cast our minds back just a couple of years: 2016 was a truly exceptional year, at least from a mathemat-ical perspective. We witnessed the political earthquake of Brexit (5/1 odds), the shock election of Donald Trump (150/1 at one stage) and the unbelievable Premier League football victory of Leicester City (5,000/1 at the start of the season). This was so inconceivable that if you'd had the foresight to place a £1 accumulator bet on all 3 of these events at their theoretical best odds, you could now be sitting smoking a Cuban cigar and sipping a piña colada with the smell of fresh Caribbean golden sand at your feet, with £4.5 million lining your pockets (and change to buy a few hundred rounds of my favourite tipple, ginger beer, or a half pint of bitter).

Actually the precise figure would be £4,503,906. To calculate this, we convert the odds into probability. If you ever see a football team being rank outsiders to a win match at 5/1, what this means is that particular bookmaker is ascribing a probability of 1 in 6 of your event happening. With 150/1 and 5,000/1, we have respective probabilities of 1 in 151 and 1 in 5,001. As these are all theoretically independent probabilities (unless Donald Trump was secretly bankrolling the success of Leicester City), you multiply all the probabilities – $1/6 \times 1/151 \times 1/5{,}001$ – to get a combined probability (called an 'acca' or accumulator bet) of 1 in 4,503,906.

Despite the seemingly unlikelihood of all these events occurring, mathematics can help us understand whether these events were truly unexpected. And further, what can the probability of improbability tell us about our own lives?

We can look at the most mundane example that many of us would have come across during our school days of maths, the toss of a coin. This is something that has a 50/50 chance of landing on either outcome, a head or a tail. Maths is all about assumptions, and as some of my students would tell me, there is a chance of a coin landing on its side. But assuming that it's a fair and unbiased coin, the head and tail are equally likely to land. If I flipped it once, I may get a head, a second time, perhaps a tail. Maybe a third time, a second head.

With each flip, the odds of a head remain the same, independent of the preceding throw. However, as the number of results increases, the more likely I am going to get closer to the theoretical probability of 50/50.

This too is a demonstration of the law of large numbers. Quite simply, the more and more you do something, the more inevitable the outcome becomes, or the closer the reality of outcomes gets to the theoretical expectation. This is something that became apparent in my days as a financial markets trader at Lehman Brothers, in the heady days where stock market predictors assumed ever continuing returns (the warning that 'past returns do not predict future performance' seemed to go unheeded).

History can give us some perspective on the law of large numbers. In 1940, British mathematician John Kerrich tested this theory while he was incarcerated in a Nazi prisoner-of-war camp in Denmark. While in solitary confinement, he sat patiently and threw a coin 10,000 times (something that my young students with ever-decreasing attention spans would find hard to comprehend). Ten thousand flips (and many, many hours) later, he accumulated 5,067 heads, or 50.67%. With each flip, the average converges towards that theoretical 50%. He published this account in a short book entitled *An Experimental Introduction to the Theory of Probability*, showing empirical validity to one of the fundamental laws of probability. Empiricism from a mathematical perspective is about using evidence from experience or observation rather than theory. Kerrich was able to demonstrate that the proportion of heads recorded asymptotically approaches the theoretical value of 50%. Until the rise of computing power and the ability to run experiments digitally, his experiment was regarded as a classic study in empirical mathematics (and painstaking repetition – I once tried to flip a coin 100 times and record the result, and started losing my mind and my motor skills by the end!).

This may appear patently obvious to us now, but the law of large numbers is significant. It guarantees that, given enough time, certain things will happen, no matter what the probability is of something happening once. A glitzy casino might lose money on a single spin of the roulette wheel, or a fortunate punter may win on a slot machine jackpot. But over time, the house will always win because the probability is on their side, and the more spins of roulette, or the more coins in slot machines, the more the reality of the outcome reflects the probability. This is the power of the law of large numbers – and ironically the more people play at casinos, the more certain it is that the house will win.

*

Have you ever found yourself sitting comfortably, getting ready for a long train journey, and then find your ex happens to be on the same carriage?

Sometimes strange coincidences are more likely than you think. For Gareth Southgate's England football World Cup 2018 squad, 2 of the 23 players shared the same birthday. Defenders Kyle Walker and John Stones turned 28 and 24 respectively on 28 May 2018. This might seem an unlikely coincidence but the mathematical birthday paradox illustrates why this actually was more likely than not to occur in a squad of 23 players.

The probability of 2 people in a squad of 23 sharing the same birthday is actually 50.9%. To work out the probability of 2 people sharing the same birthday, you work out one takeaway: the probability that everyone has a unique birthday. So let's work out the probability that everyone in a group of 23 players has a unique birthday.

For player 1, it is 100% that he has a unique birthday as every date is clear. For player 2, there is 1 day that he could share, but the other 364 are free. So the second player's chance of a unique birthday is 364/365. For the third player, it's 363/365. We can continue this pattern till the twenty-third player, which is 343/365.

To calculate the probability of every player in the 23-man squad having a unique birthday, we multiply all these probabilities together. So $365/365 \times 364/365 \times 363/365 \times \ldots \times 345/365 \times 344/365 \times 343/365 = 0.491$. So the chance of a shared birthday is $1 - 0.491 = 0.509$ or 50.9%. (Note that we have ignored leap years in our problem.)

I have personally had a seemingly unthinkable event happen to me. During the Christmas holidays of 2014, I was taking a short break during my teacher training year at Hughes Hall in Cambridge. My cousins and siblings frequently venture on short breaks together – often a short Ryanair flight away to somewhere in Europe – but on this occasion, we had ventured to the west coast of America. Staying in the casino capital of the world, glamorous Las Vegas, we embarked on a day trip to the Grand Canyon by the Colorado River. As a child, I remember watching re-runs of the illusionist David Copperfield apparently levitating across the Grand Canyon. So to have had a helicopter ride across the canyon was something of a childhood dream come true.

Here I am, thousands of miles away from my life back in the UK, and I feel a shoulder tap accompanied by a British accented voice saying, 'Is that you Bobby?' Turning around, much to my astonishment (and that of my family), there was a member of staff from my college in Cambridge. Of all the places in the world I could have been, she happened to be in America, in the Grand Canyon and on the helicopter next to us! What were the odds? A miracle, some might say.

Again, a mathematical perspective can help us understand why this might happen. Let's go to my university town of Cambridge and look back at the hippy days of the 1960s. In 1968, Cambridge professor John Littlewood theorised that due to the Law of Large Numbers, given any big enough sample, you could expect any outrageous thing to happen. Littlewood's law states that a person can expect to experience an event with the odds of 1 in a million at a rate of once per month. He defines a 'miracle' as something with a 1 in a million chance of occurring.

Littlewood's theory appears straightforward. During the 8 hours we're awake and alert in a day, we see and hear things at a rate of once per second. A basic bit of maths (8 hours × 60 minutes × 60 seconds) gives us 28,800 possible events. So roughly, over the course of a day, we must witness close to 30,000 different things. Most are completely mundane (such as seeing a traffic light go green or hearing a bird cuckoo in the morning), but over the course of just 35 days (34.72 to be precise), we'll have witnessed more than a million possible events (again the maths behind this is 1 million divided by 28,800).

Therefore a million-to-1 event will happen to every one of us, on average, every single month. With a global population of more than 7.6 billion people, highly unlikely events are certain to happen every day to people all around the globe. It's the inevitability of improbability. We only notice the highly extraordinary and seem to forgot about the plain ordinary.

It is this reason why gamblers might remember their lucky wins and forget their unlucky losses, something that I was guilty of as a professional trader (and in my gambles as a sports fanatic and long-suffering

West Ham fan). The same rationale applies to why psychics can make so many predictions in the hope you remember the ones that hit, as opposed to the ones that miss by some margin.

So if you feel like it hasn't been your day, your week or even your year, just blame it on the inevitability of improbability! However, be careful not to fall into the trap of the gambler's fallacy. Have you ever been to a casino and seen the roulette wheel land on red 4 times consecutively, and thought to yourself that the black is long overdue? If so, you have fallen prey to the mathematically erroneous belief that if an event occurs more frequently than normal during a particular period, it will happen less frequently going forward. If we assume that the outcome is truly random (and that the casino doesn't have a crooked way of fleecing you!) and the wheel's outcomes consist of independent trials of a random process, a lucky streak of 4 reds does not mean a black is about to appear. In the probability of independent events, future events are unaffected by previous flips.

The gambler's fallacy had one of its earliest written descriptions by French mathematician Pierre-Simon Laplace in 1796. As well as being an expert in astronomy (he was one of the first scientists to postulate on the existence of black holes), he is often referred to as the Isaac Newton of France. In 'A Philosophical Essay on Probabilities', he described the ways in which men calculated the probability of their wives giving birth to sons:

I have seen men, ardently wanting a son, who could learn only with anxiety of the birth of boys in the month when they expected to become fathers. Imagining that the ratio of these births to those of girls ought to be the same at the end of each month, they judged that the boys already born would make more likely the births next of girls.

Laplace was explaining how expectant fathers feared that if other men were having sons, it meant they were more likely to have a daughter to balance out the probability. And herein is the gambler's fallacy. Each birth is independent of the other, and if the probability of a boy is 50%, then it shall be so irrespective of that particular month's count at the local hospital or indeed any other births ever.

So the law of large numbers give us a compass to navigate the randomness around us. It states as the number of trials or observations increases, the actual probability approaches the theoretical or expected probability. If you just focus on the probability of a single event occurring, many things can seem vanishingly unlikely, with a probability close to 0 and therefore a trivial chance of happening. But you need to know how many times that experiment will be run, because with enough time, enough spins of the wheel, the probability of unlikely events becomes increasingly likely.

Rather than giving up on things that probably won't happen – like winning a prize in a competition – I try to see life through a more optimistic lens. Some may say it's blind optimism, but I prefer the words of Kipling in his poem 'If', where the narrator advises perseverance even:

when there is nothing in you
Except the Will which says to them: 'Hold on!'

A poster of this poem has been in my bedroom for a couple of decades and has reflected my belief that better fortunes might be waiting around the corner for us (my football club West Ham's anthem 'I'm forever blowing bubbles' talks about 'how fortune's hiding' and you've 'looked everywhere'). I believe that we can keep looking till we find what we seek. Life can be challenging or throw obstacles at us. But I would like to think that the more you try at something, the more likely you'll succeed. With the law of large numbers, it turns out there's actually a lot of method behind the madness of the phrase 'If it at first you don't succeed, try, try, again.'

Chapter 2 Puzzle

Happy Chinese New Year

For the Chinese new year, you win a competition to attend a party with lots of animals. In 2018, we celebrated the year of the dog, taking over from the rooster of 2017 and the monkey of 2016.

The party is being held at the zoo and the top dogs, roosters and monkeys are invited.

There are twice as many dogs invited as roosters, and twice as many roosters invited as monkeys.

Assuming all the dogs have 4 feet and the roosters and monkeys have 2 feet, if there are 88 animal feet at this party, how many dogs, roosters and monkey are there? Do not count your own legs!

CHAPTER 3

'Collect, Collect and Collect Some More'

The Numbers Underpinning the Hobby of Collecting

Growing up in the state of Kerala in southern India, my father used to be an amateur phillumenist. 'What is this?' you say. It derives from the Greek word 'phil' for loving and the Latin 'lumen' for light. A person who engages in phillumeny partakes in the hobby of collecting match-related items, in my father's case, matchboxes. As well as collecting matchboxes, my father had small-scale collections of other items: pebbles, seashells and marbles.

One is not born but rather becomes a collector (with apologies to the French intellectual Simone de Beauvoir). Although of course we can't inherit hobbies from parents genetically, they are often something that can be instilled in children. My father encouraged me and my elder brother to throw ourselves into collecting. As children, we are often naturally drawn to accumulating or hoarding things. Some find joy in miniature toy cars or natural objects such as stones, pine cones or shells. Others collect simple household objects such as assortments of fabrics or buttons.

My father started me off with some coins and stamps, giving me the base to practise amateur numismatics (coin collecting) and philately (stamp collecting). Our home town of East Ham is chock-a-block with delightful charity shops and provided plenty of rich pickings for our coin- and stamp-hunting. Likewise, we would implore relatives from home and across the world to post us any weird and wonderful coins or stamps that crossed their path.

Stamp collecting allowed me to think about different ways of categorising things. To the untrained observer, stamps are all just the same: a small adhesive piece of paper that is stuck onto the front of a letter to show that a certain amount of money has been paid. But once you start collecting them, you soon discover various ways of organising and ordering them to make sense of your collection. Sizes, country of origin, condition, colours, age, mint or used – these are just some of the most obvious categories. Helping children learn to compare and describe items that from the outside look very similar allows them to understand different ways of measuring. Spotting patterns in my stamp collection – turning them into a kind of dataset – helped me learn how different numbers connect to each other.

The pinnacle of my stamp collecting occurred on a warm Saturday, 29 April 1995. I distinctly remember our

London tube journey that day, overflowing with exuberant rugby league fans covered in the yellow and blue kits of Leeds and the cherry red and white of Wigan (another example of categorising different fans). It was the day of the Challenge Cup final, rugby league's oldest knockout competition. Flags were fluttering proudly down Wembley Way, the road that linked the local tube station and Wembley Stadium. My story does not end up in the famous national stadium, but in the nearby Wembley Exhibition Centre for I was on my way to 'Stamp 1995', a 4-day exhibition devoted to all things stamp related.

As a child of 11, I was in paradise. I wandered through rows and rows of stamps from all around the planet, and heard exhibitors and punters tell outlandish tales of how they came to acquire them. More recently I've been to other marvellous conventions such as Comic-Con, stacked with visitors attired in splendid superhero outfits, but nothing like this was ever quite as exciting to me as Stamp 1995. And the highlight of the day was when we purchased a damaged but affordable Penny Red, with Queen Victoria's head in stately profile, which succeeded the Penny Black, the first adhesive stamp used in a public postal system.

The stamp was reverently escorted back to East Ham and securely placed in my stamp book. Unfortunately, all these years later we're not quite sure where this book has disappeared to (not an easy task to locate as our house is rammed with thousands of books) but I still hope that, one day, we might stumble upon the stamp that could be worth up to a few hundred pounds!

So why do we collect things? Is it just something to do? Or a way of saving things that will accumulate in value over time? It's not just something for children. TV show host Jonathan Ross collects rare comic books. Actress Angelina Jolie collects knives. Rocker Rod Stewart collects model railways. Actor Johnny Depp, special edition Barbie dolls. Tom Hanks, typewriters. The collecting bug can be caught by small boys in their bedrooms, or by some of the most rich and famous people in the world.

In the UK, about a third of people collect something. Our desire to collect was only possible as a species after we abandoned our nomadic ways, and started settling down, around 12,000 years ago. It appears a uniquely human pastime. And once we collect a significant number of whatever takes our fancy, the next step is to order and categorise what we have. There is also a psychological phenomenon at work known as the endowment effect, which argues that we tend to value things more once we own them – it seems we're more worried about losing what we have than excited about gaining something new.

As I mentioned earlier, some of my most prized possessions are my (nearly) completed Premier League sticker books from the mid-1990s. These probably wouldn't garner much re-sale value as they are not that rare, but they are priceless to me (in other words, they hold sentimental value). To most sensible adults, the word *panini* conjures up the image and smells of a freshly grilled Italian sandwich. But mention the word

Panini to children, especially those into football, and their first thought will be about collecting stickers. (If football isn't quite your thing, Panini offer sticker collections for all types of interests: Spider-Man, My Little Pony, the Power Rangers, Thomas the Tank Engine and even publish the official *Doctor Who Magazine*).

In the run up to major international football tournaments, such as the European Championship or the World Cup, the machines at Panini's production headquarters work overtime. The automatic packaging machines known as 'Fifimatics' work tirelessly for 21 hours a day, 6 days a week, packing around 8 million stickers bearing the photos of hundreds of players.

With a bit of back-of-the-envelope calculation, we can work out the total available sales that the company could make based on UK prices. For the most recent 2018 World Cup, a packet had 5 stickers and cost 80p. So with 8 million stickers divided by 5 stickers, we have 1.6 million packets. At 80p per packet, this gives a total sales level of £1.28 million. Not bad for just selling photos of football players printed on adhesive labels!

But as well as the total available sales for Panini, we can also use mathematics to work out the cost of acquiring the stickers as a consumer. And be warned, the amounts are not pretty! So if you a child trying to convince your parents to fund your sticker habit, I urge you to hide the next couple of pages of the book unless you want to either give them a heart attack or for them to urge you to find a more affordable hobby – maybe collecting stones or something free!

OK, here we go.

There were 32 squads in the 2018 World Cup. For the purposes of the sticker book, each squad has 18 players (real squads have 23 players). Additionally, there is a sticker for the entire team and the country's logo, so in total there are 20 stickers per country. So $20 \times 32 = 640$ so far. But there are also 32 miscellaneous World Cup stickers, for the different stadiums in Russia and other World Cup-related items such as the trophy and the ball. So $640 + 32 = 672$. Then we have a final 10 stickers for legends of the game and the most iconic teams of all time, taking us up to a grand total of 682 stickers.

So with a sticker packet costing 80p for a set of 5, it costs 16p per individual sticker. So if we could buy 682 individual stickers, we would get $682 \times £0.16 = £109.12$. However, as sticker packets come in packs of 5, the least number of stickers you can buy needs to be divisible by 5. So you'd have to buy 685 stickers, which is 137 packets and a total cost of £109.60.

But if you have ever purchased stickers before, you will know that it is virtually impossible to buy 137 packets without getting repeats. If you know anyone this fortunate, I suggest you ask them to tell you next Saturday's lottery results, pick a winner at the next Grand National horse race and predict the next Bitcoin craze for you! But £109.60 is our mathematical minimum assuming no repeats.

However, the real question for the typical collector is how many stickers on average you have to buy to complete a collection. At first it may seem impossible to know but, with some clever maths, we can use a probability model to work it out. Our saviour is an adapted version of the 'coupon collector's problem' which we shall call the 'sticker collector's problem'. This helps to describe the 'collect all stickers and win' contests. In this problem, our aim is to purchase distinct objects so as to have a complete set of objects. Each purchase gives us a random object, and each separate purchase is independent of the previous ones. In our case here, the target is the eternal glory of a completely filled-out sticker book. Mathematically speaking, our aim is to calculate the probability of buying a sticker we don't already have. As we accumulate more and more stickers, the number of stickers we need gets smaller and smaller, and thus the probability of getting that sticker in our next pack gets smaller and smaller too.

So the first sticker we buy is cast iron guaranteed not to be a sticker we already have, as we don't have any stickers. When we acquire our second sticker, this has a 681/682 chance (99.85%) of being a sticker we don't have in our collection. Now we have a second unique sticker, we can calculate the probability of the third sticker being unique as 680/682 (99.71%) and this pattern continues.

If we continued this calculation down to the final sticker – so $682/682 \times 681/682 \times 680/682 \times 679/682 \times \ldots$ $3/682 \times 2/682 \times 1/682$ – this will work out the probability of us never getting a duplicate on our way to completing the collection. The numerator of these fractions – the number *above* the line in the fraction – reduces by one each time: $682 \times 681 \times 680 \times \ldots 3 \times 2 \times 1$.

This is called a factorial. A factorial is a function that multiplies a number by every number below it. Factorial 5 is written as 5! (or 5 shriek as some funky mathematicians might call it), and it's calculated as $5 \times 4 \times 3 \times 2 \times 1$ (the answer is 120). Our numerator of the fraction for this problem is 682! Our denominator – the number *below* the line in the fraction – is $682 \times 682 \times 682 \times \ldots \times 682$ but this happens 682 times. If we did $4 \times 4 \times 4 \times 4 = 256$, we could write this as 4^4, read out as '4 to the power of 4'. If we did $7 \times 7 \times 7 \times 7 \times 7 \times 7 \times 7 = 823{,}543$, we could write that as 7^7. Hence our denominator is 682^{682}. So the calculation would be 682! (682 factorial) divided by 682^{682} (682 to the power of 682). The denominator is much, much, much larger than the numerator so we would have an incredibly small fraction in the end. I've tried using online calculators to get an answer for this, but the number is far too close to 0 to get an answer I can write down here. So mathematically, are you going to able to buy 682 stickers without a single repeat? Nope. Not going to happen, not in this universe.

We don't want to work out the probability of each sticker being unique, but what we'd like to calculate is how many stickers we should expect to buy each time for the next unique sticker. The maths is going to get a little involved, so buckle up.

Let's say that the probability of an event occurring is 'p', then the expected number of times we have to do something to get that outcome is $1/p$. For example, if the probability of West Ham's skipper Mark Noble scoring is 0.1, then the number of games he would have to play on average to score is 10 (which is 1 divided by 0.1). With our sticker collection, we need to calculate the following sum: $682/682 + 682/681 + 682/680 + \ldots + 682/3 + 682/2 + 682/1$. In maths, we call this a harmonic series. We can use a mathematical estimate to work out where this sum will end up: $n\,(\ln(n) + y)$

ln = natural logarithm (you'll see this button on your regular scientific calculator)

y = the Euler-Mascheroni constant which is rounded to 0.577

n = the number of stickers you are required to collect

As there are 682 stickers in the 2018 World Cup collection we have n = 682. Plug this into our equation and we get:

$682 \times (\ln(682) + 0.577) = 682 \times 7.1 = 4{,}844$ (using rounded figures)

So on average, we would expect to have to buy 4,844 stickers to complete this album. This works out as 969 packets of 5 stickers at a sum of £775.20. This is more than 7 times the cost than if we had no duplicate stickers.

If you're on the ball, you will have noticed the flaw with this initial calculation. Stickers are not purchased individually but come in those packets of 5. Panini promises each individual packet does not contain duplicates.

So our calculation is a bit more refined to:

$$[(682/682) + (682/682) + (682/682) + (682/682) + (682/682) + (682/677) + (682/677) + (682/677) + (682/677) + (682/677) + \ldots + (682/2) + (682/2) + (682/2) + (682/2) + (682/2)]$$

(In the denominator, you will notice that there are 5 lots of 682 and then 5 lots of 677, which refers to the fact that packets come in 5s.)

$$= 682 [5/682 + 5/677 + 5/672 + \ldots + 2/5]$$
$$= 682 [1 + 1/2 + 1/3 + \ldots + 1/136] \text{ (I've slightly simplified the calculation here ...)}$$

Here we simplify the equation using the method of natural logarithms, which is the ln.

$$= 682 [\ln(136) + y]$$
$$= 3,744 \text{ stickers which is 749 packets at a total of £599.20}$$

We have used a mathematical model to help us out and, in mathematics, we often make assumptions to make our calculations more manageable – there is a trade-off between precision and the economy of efficient calculations. We assume that every sticker has exactly the same chance as appearing as another one, which in the real world assumes that Panini prints the same amount of each sticker and distributes them randomly. Whether they produce the same number of elite players such as Portugal's Ronaldo or Argentina's Messi is another thing.

Further, we are talking about buying packets of stickers one at a time. There are offers available on Amazon where you can buy discounted boxes of 100 packets at a time (imagine trying to open 100 packets in one sitting!). In the real world, it is unlikely that sticker packet collectors will buy in a vacuum, with no contact with anyone else. You will often trade stickers with friends and this will reduce the number of packets you need as well.

Alternatively, there is a corner-cutting method for sticker fans who appreciate searing efficiency and want to complete their collection at the lowest cost. Panini allows individuals to buy up to 50 stickers for 'customers who wish to complete their own collections only'. Should you be so unscrupulous as to recruit several friends or family with different house addresses, you could complete the collection fairly painlessly. With 682 stickers, you need 13 other willing participants to order 50 stickers on your behalf. At 22p per sticker, $682 \times £0.22$ gives us £150.04 to complete the entire collection (excluding shipping costs).

So what? Now if you are ever going to embark on a collection of a finite set of items that come out randomly like stickers, you are (probably) able to use a probability model that will help you to establish how much you have to spend! And once you've done that, tuck that collection away in an attic to collect dust. Dust that could build up for many decades.

In March 2018, ahead of the build-up to the Russia 2018 World Cup, a Leeds United fan from Wakefield, Jonathan Ward, sold his 1970 Panini Mexico World Cup sticker book for a staggering £1,550 on eBay! Not bad considering how little he spent on it (filling it would have cost a few pounds in total back then). Further consider the wear and tear the book had suffered, the 6 missing stickers and even some scrawling from his 9-year-old self ('Leeds are the greatest' next to Jack Charlton's name). So hardly mint condition, but Mr Ward was able to convert his fortuitous attic rediscovery to cold cash.

A cynic is 'a man who knows the price of everything and the value of nothing'. These are the words of Lord Darlington in Oscar Wilde's 1892 comedy *Lady Windermere's Fan*. From my days as a chartered accountant at PwC, some might aim this accusation at me! My job as an accountant was to ascertain whether the price of items on a company's balance sheet was fair. At its most basic level, junior accountants were often involved in 'stock takes', where you would physically verify the quantity and condition of items held in inventory or a warehouse.

Collecting as a child – be it stamps, coins or stickers – allowed me to develop a sense of how to organise using categories. Mathematics is often about pattern spotting and being able to see where things fit in the big picture. Even my knackered music CD collection, ordered at one stage alphabetically and then, for more aesthetic reasons, by the colour of the external CD case, can demonstrate this. I think starting collections can be an entertaining method to get young people thinking about numbers as they make decisions about why certain objects should be ordered in a particular way.

The nature of our relationship with numbers can be thought of using an analogy from the Singaporean method of teaching maths. They start off with the 'concrete', items you can physically grasp, such as toy cars. Children are able to count what these items are: 1 car, 2 cars, 3 cars and so on. Once this is embedded, children progress onto the 'pictorial' representation of toy cars. They understand that there is a distinction between a diagram of 1 car and 2 cars. Then finally, the breakthrough to that 'abstract' level of thinking where children are able to discern that 1 car can be represented by the number 1 and 2 cars with 2. From there, children are able to steam ahead with mathematical understanding – adding, subtracting, multiplying, dividing with these numbers. Once children are comfortable thinking about numbers abstractly, there are no limits – 'to infinity and beyond' to borrow a phrase from Buzz Lightyear from the *Toy Story* movie series.

But to get there, children first need to master counting concrete objects, things they can physically hold. As the world gets increasingly digital, children have less exposure to physical objects. CDs, coins, stamps or building blocks are all less common these days, and my collections of coins, stamps and even stickers may become relics of the past. These days it's nostalgic adults who are the biggest sticker collectors, and many young children are much more enamoured with the latest app on their smartphone than purchasing expensive bits of adhesive paper.

Between school and university, I took a gap year. The first 9 months were spent at another Big 4 accounting firm, KPMG, on their gap-year scheme. Most of my time was insightful and productive, but I spent one week in a windowless storage room with masses of filing cabinets, reorganising folders into a particular type of order. To be honest, I quite enjoyed the monotony for the first few days but I ended up feeling like the goblins eternally shifting items in Gringotts' underground vaults in Harry Potter's Diagon Alley by the end. But I wonder if we need the physicality of objects for us to start thinking about relationships between different items? The future is set to become increasingly digitised. Having a perfectly arranged e-mail inbox or Spotify music playlist doesn't quite have the same emotional appeal for me as a beautifully ordered CD collection, and I worry the next generation may not learn the value of categories quite so easily. But the future is digital and perhaps the hobby of collecting may soon be consigned to a far-flung corner of the British Museum.

Chapter 3 Puzzle

Stamp-collecting Holiday

A logically minded stamp collector is planning her summer holidays and is thinking of going to the following countries in this particular order: Jordan, Algeria, Ethiopia, Ghana and Nigeria. Which country should this stamp collector go to after these to maintain her attention to detail in getting stamps in a particular order?

CHAPTER 4

'It's a Kind of Magic'

The Maths Behind Magic

Are you sitting comfortably? Then I'll begin. I want you, my trusted reader, to follow me. So read this paragraph carefully and get ready to be (mildly) impressed, I hope.

I'd like you think of a number from 1 to 10. When you have your number, multiply it by 9. Now, if this is a 2-digit number, add the 2 digits together. If it's not a 2-digit number, do nothing. Still with me? OK, now subtract 5 from the number in your head. What I want you to do now is to think of a letter in the alphabet that corresponds to the number you are thinking about. For example, if you have a 1, then that is an A, if you have a 2, then that is a B, and so on. I want you to think of a country's name that begins with the letter you are thinking of. Now with this country, think about the last letter in that country's name. Now, without hesitation, think of an animal whose name begins with that letter. Now, think about the last letter in that animal's name. Think of a colour that begins with that last letter in the animal's name. And finally, put the colour animal and country together. And drum roll … What are you thinking of?

An orange kangaroo in Denmark? (Hopefully this is what you got, rather than a turquoise ant in Dominica or aqua iguana in Djibouti!)

My family used to attend magic fairs and exhibitions in the early 1990s and I remember once following this riddle. Being led like a young horse in a show jumping competition through the hoops, I experienced a pure delight that perhaps only children (or the naïve!) can, when the host asked who was thinking of orange kangaroos in Denmark. After that initial amazement passed, I was keen to understand how the magician had led me to say what he wanted. Was I really a victim of mind reading or was something else at play?

Upon further investigation, I realised that some basic maths combine with assumptions about our most likely hunches to create this impressive-sounding party trick. Any single digit number multiplied by 9 will always yield an answer of 9 if we add its 2 separate digits (look at a list of the 9-times tables to see this: 9, 18, 27, 36, 45, 54, 63, 72, 81, 90 – these digits always add up to 9). Subtracting 5 from 9 gives us 4, which equates to a 'd'. And most of us will pick Denmark as the first country beginning with 'd' that comes to mind. And if we choose Denmark, and are asked to pick an animal beginning with the last letter 'k', most people will opt for a kangaroo (though you may very occasionally be caught out with a kiwi or a koala). And finally, in choosing a colour beginning with 'o', most people will pick 'orange' and thus you end up with the orange kangaroo in Denmark. There are variations of this maths trick that gives the answer of elephants in Denmark, if I had asked you for the name of an animal that begins with the second letter in the country's name, 'e'. (Curious trivia alert: there is an actual Order of the Elephant in Denmark, which is Denmark's highest ranked honour, almost exclusively for royalty and heads of state.)

Growing up as a child in the 1990s, I was obsessed by the Sony Walkman, fluorescent bumbags and a glut of magic shows on TV. British television was inundated with prime-time shows that centred on magic.

Paul Daniels, and his assistant the lovely Debbie McGee, was a mainstay of BBC1 in the 1980s and early 1990s, his show regularly attracting 15 million viewers at its peak. Internationally, the megastar illusionist David Copperfield, described by *Forbes* magazine as the most commercially successful magician in history, was also frequently on British telly. (And yes, he did actually take his stage name from the famous 1850 Charles Dickens novel, in case you were wondering.) Magic was everywhere at that time and these programmes made great family viewing, for adults and kids to enjoy together. However, the TV bosses soon cottoned on to how much kids loved magic, and started to make bespoke magic shows specifically for children. I would rush back from school to watch *Tricky Business*, and enjoyed the explanations behind the tricks almost as much as the initial wonder of the 'magic' itself. In an era before you could search online for the secrets behind tricks in an instant, you could allow yourself to linger longer in that moment of initial wonder and enchantment when watching a magic trick on TV. How do they do it?

If you ask a child now about magic, older kids might tell you about the levitational powers of the *wingardium leviosa* spells in Harry Potter and younger ones talk about Elsa's powers from Disney's *Frozen*. Adults might more often mention international performers such as David Blaine or Derren Brown. Much magic that we see on screen or stage relies on props or sleight of hand, but there is a very particular type of magic that relies on maths. A magic that relies on logic, and as someone who loves maths and numbers, this is the kind of magic that I have always found most appealing. Why? Because you can figure it out yourself, and once you get it, you can do the trick yourself.

Time for a quick quiz question. Ready? 'Any sufficiently advanced technology is indistinguishable from magic.' This is the third of 3 eponymous laws named after which British science fiction writer, most renowned for co-writing the screenplay for the 1968 film *2001: A Space Odyssey*. Got it? It's Arthur C. Clarke. It is one of my favourite quotes and no matter how much technology advances, it's always true. To people even a few generations ago, the thought that we can now communicate across the world instantly via face calls would have felt like magic to them!

I mention this quote now because to some extent, when I was a child and was mesmerised by magic tricks, the sense of wonder came because I could not understand the technology behind it (be it the basic algorithm or perhaps a sleight of hand or an optical illusion). That initial 'wow' comes before the mind begins attempting to process how logic could underpin a trick.

In my high street, there are often street hustlers with a ball hidden under 1 of 3 cups, trying to persuade the public to part with £20 for a chance to double their stake by finding the ball. It can seem magic, but their trade relies on sleight of hand combined with some patter and misdirection. You might think only suckers or simpletons would try to take these guys on, but even the most towering of intellects can lose their footing for a moment when encountering a magician. Albert Einstein is perhaps the most iconic 'genius' in history – when my school students complain that a particular maths topic is beyond their reach, I often refer to Albert's quote: 'Do not worry too much about your difficulties in mathematics, I can assure you that mine are still greater.'

But on one occasion came the infamous 'Trick That Fooled Einstein', performed by the British magician Al Koran (real name Edward Doe). Unfortunately there are no video records of this interaction but it probably went something like this (and I've exchanged the original dollars for pounds as my brain works that way):

> Al Koran: 'What would you say if I told you, Albert, that I could guess the amount of change in your pocket?'
>
> Albert: 'Well, I would be say that is scientifically not possible.'
>
> Al Koran: 'OK, I'm going to do more than that, I'm going to make 3 predictions. I have as much change in my pocket as you do, plus an extra 50p and then enough remaining change to bring your total to £2.35.'
>
> Albert: 'OK, £2.35, that's what you say.'
>
> Al Koran: 'So place your loose change on the table. I'll do the same thing with my change. I have £2.85, how much do you have?'
>
> Albert: 'I have £1.71.'
>
> Al Koran: 'I will remove £1.71 from my pile and place it next to yours. Remember I said that I have as much as you and an additional 50p. So I'll take 50p and move it to my pile. So I have 64p left, which is £2.85 less the £1.71 and less the 50p. Still with me, Albie?'
>
> Albert: 'Yes, still with you.'
>
> Al Koran: 'Finally, I said that there would be enough left over in my pile to bring your total to £2.35. So what'll I'll do is move my 64p into your pile. Please could you now add up your pile?'
>
> Albert: '£1.71 + 64p gives a total of £2.35. And this was your prediction!'

Albert then proceeded to ask Al Koran to repeat this trick, and the great physicist was still dumbfounded. 'It's not the numbers, but the words that fooled you,' were Al Koran's concluding remarks. This is a case of Einstein perhaps being too smart for his own good. He may have assumed that complex maths was at work, when it was actually just smoke and mirrors in the form of a simple language trick. I'll reveal exactly where Einstein went wrong, so listen carefully.

The secret is just the wording. At the start, Al Koran said that he was going to guess the amount of change Albert was carrying, but that was not what he ended up doing, even if it appeared so. Al Koran then proceeded to claim that he would deliver even more, that he would end up making 3 predictions. The reality is that the 3 predictions when put together were logically equivalent to saying 'I have £2.85', which is not that revealing. Al Koran began with £2.85 and removed an amount equal to X from Albert's pile, along with 50p, leaving Albert with $£2.85 - 0.50 - X = £2.35 - X$. We then took this remainder and added to Albert's pile of X. The maths is essentially $£2.35 - X + X = £2.35$. It's not a maths trick, just a language one!

Albert got caught up listening to the story that Al Koran was weaving, and missed the logic. It demonstrates to me what a lot of powerful magic is about. Magic tricks always have some element of deceit, be it physical or mathematical, but it is the story the magician weaves that keeps us engaged and perhaps makes us miss the obvious.

This neatly ties in with an explanation of the 2 types of reasoning our brain uses. One of the books my father introduced me to recently was the Nobel Prize- winning economist Daniel Kahneman's 2011 book, *Thinking, Fast and Slow*. The central thesis of the book is the interaction between 2 modes of thought. System 1 thinking is fast, instinctive and emotional. System 2 thinking is slower, more deliberate and more logical. All of our decisions are a kind of battle between intuitive and analytical thinking. Our mathematical thinking lies in System 2, but System 1 will often break through when we are distracted or misguided. Even mathematicians have to suppress their intuitive System 1 processes at times.

A classic puzzle question demonstrates this tension between System 1 and System 2 thinking really clearly:

A bat and ball cost £1.10.
The bat costs £1 more than the ball.
How much does the ball cost?

What does your mind instinctively say? If you're like me, the answer that jumped out when I first heard this puzzle was 10p. But that's not correct. What makes us go for this wrong answer is that it 'evokes an answer that is intuitive, appealing and wrong', according to Kahneman. If it was 10p for the ball, then the bat would cost a pound more at £1.10 and the total would be £1.20. The actual answer is £0.05 and £1.05. More than 50% of students surveyed at the top American universities (Harvard, MIT and Princeton) gave the intuitive but incorrect answer. And magicians, especially those using maths in their tricks, can exploit this by trying to direct our attention to what feels right, rather than what is logically correct.

David Copperfield is famous for his mind-reading tricks, and one of the most famous ones uses a simplified version of a complex mathematical technique called the Kruskal Count, devised by American mathematician and physicist Martin Kruskal. In Copperfield's simplified version of the trick, Copperfield has a circle, the numbers 1–12 like a clock, and participants choose a number to start with. Following a set of procedures, every one of us ends up with the 6, which seems miraculous. But it relies on maths – in this case on the algorithmic Kruskal Count. I had actually forgotten about this trick for years until it was a Starter for 10 question on the 2017–18 series of *University Challenge*, and it compelled me to look it up on YouTube. I would encourage you do the same and be dazzled by David Copperfield's mind reading (search for 'Magic of David Copperfield in your own home'). His power of mind reading still works 20 years later online!

But there is much more to maths and magic than simplified algorithms. One of my favourite maths tricks was devised by Dutchman Nicolaas Govert de Bruijn. Picture the following scenario. A mathematician hands out a pack of cards to a volunteer, who will repeatedly cut the deck. The mathematician will then hand out the first 4 cards to 4 volunteers. Using apparent telepathy, the mathematician needs to ask a few questions to his volunteers to make sure his mind-reading skills are functioning. 'Who ate cornflakes for breakfast?' 'Is anyone an Aquarius?' 'Does anyone support West Ham?' Who is holding a red card?' Just with this information, the mathematician is able to name the card each person is holding, hopefully to rapturous applause from the audience.

Now the secret behind this is mathematics, as expected. You may have noticed that 3 of those questions were red herrings, it was only the question about the red card that held any information. The information was encoded using a de Bruijn (it's pronounced de Brown) sequence which is an order of 16 cards in this exact order BBBB RRRR BRRB BRBR (with B and R for black and red). You do this for 3 lots of 16 to give the appearance of a pack of cards (48 is close to 52).

There are 16 ways to arrange 4 red and black cards. This sequence contains every combination of red and black cards. So if you wanted a red red black red, or a black red black red, you can find that somewhere in the sequence. As soon as one volunteer tells you where the red cards are, you can work out exactly where you are in that sequence. And here's the tricky part: if you've memorised the sequence, you can then reel off the cards chosen by the volunteers. On a more serious note, de Bruijn sequences have far-ranging applications beyond magic, and are used in fields from robotics to neuroscience, safe-cracking and even DNA sequencing.

So how can magic play a role in exciting people about mathematics these days? If I mention the word busking, you wouldn't naturally think of maths. We often expect buskers to be performers who play music, dance or even perform a comedy act. But a few years back I was enchanted by Dr Sara Santos doing 'maths busking' on the BBC. Sara is a charismatic performer, wearing a bright yellow hat as she ties up people with ropes and guesses their birth dates, all using straight-forward maths. Performers like Sara are a fabulous way to get people engaged with the power of numbers and to show people how maths can be fun and surprising, as well as useful.

In a world where numbers sometimes feel like dry data, magic – using maths – still has the power to enthral and give us a sense of innocent wonder. If maths and magic can play a small role in that, I'm all for us learning a few maths tricks.

Chapter 4 Puzzle

Harry Potter and the Quest for 4 Pints of Butterbeer

Harry Potter, Hermione Granger and Ron Weasley head off after their OWL (Ordinary Wizarding Levels) exams results at Hogwarts School of Witchcraft and Wizardry to celebrate with 4 pints of Butterbeer at Hogsmeade Village. To show camaraderie, they will all be drinking from the same tankard. They visit The Three Broomsticks pub but find that the bartender has run out of clean 4-pint tankards.

Ron immediately gets his wand out to perform magic to clean the 4-pint tankards but Harry reminds Ron that they promised not to use magic as they were celebrating. Hermione then claimed that she could use maths, and not magic, in order to create 4 pints of Butterbeer. She has found an empty 5-pint tankard and an empty 3-pint tankard.

With only 10 pints of Butterbeer (and no magic!), how does Hermione create 4 pints of Butterbeer?

CHAPTER 5

'To Boldly Go Where No Math Has Gone Before'

The Maths of Space

Mercury, Venus, Earth, Mars, Jupiter, Saturn, Neptune, Pluto. My primary school, St Michael's in East Ham, used the name of planets in order from the Sun for the names of our classes. Any observant reader will point out the absence of one planet, which was the first one discovered with a telescope in 1781 by William Herschel. Indeed this planet was due to be called Georgium Sidus (George's Star) in honour of Herschel's patron, King George III. Final clue: in one episode of the cult cartoon series *The Simpsons*, Bart bursts into the living room with a water gun, a red helmet with antennae popping from it and green googly eyes and shouts 'I am the thing from … [planet]!' Well, it is Uranus! I never officially found out why my primary school didn't have Uranus as a class, but it's safe to assume that children aged 5–11 would not be able to resist chuckling at the class name (for obvious reasons). Looking at my primary school class list, Pluto has sadly been demoted by the International Astronomical Union from planet status to 'dwarf' planet status.

As a child, my earliest ambition was to be a cosmonaut. Yes, I know it should be an astronaut, but the influence of Russian astronomy books from my parents' homeland in Kerala in India had made me think the Anglo-Saxon word was cosmonaut. I'll be honest, I was never the largest fan of fairground rollercoasters that would hurtle through the air at seemingly astronomic speeds, but I would go on them purely to prove a point to myself that I could one day withstand cosmonaut training. However, I remember once hearing from a friend at school that I would be too short to be an astronaut, and unfortunately I think I may have let this dissuade me from this dream! If only Google had existed, I would have found out that organisations such as NASA are moderately generous for those of us who are vertically challenged. Apparently to be a commander or pilot astronaut, you have to be between 158 and 190 centimetres, but if you're happy just to be a mission specialist, then you can be as short as 149 centimetres. This didn't curtail my sense of amazement when thinking about the universe.

The world of mathematics and physics can throw up some results that initially seem counter-intuitive to the layman. I remember aged 15 or 16, during my Sixth Form Scholarship interviews to Eton College, I was in a Physics interview. I handled questions on the transmission of electricity comfortably but then I was slightly thrown by one question. *If a person jumps from the top of the school library* (which looked a bit like a mini-version of London's St Paul's Cathedral to my uninitiated eye), *what physics goes on?* Most GCSE students would describe the fall, perhaps raise the possibility of terminal velocity from taller buildings and air resistance. But after some prodding, I eventually got to what the interviewer was looking for – Newton's Third Law of Motion: for every action, there is an equal and opposite reaction. If a person (an object for the purposes of this interview) were falling towards the Earth, the Earth would move upwards in response. By how much? Such an incredibly small amount, but it was something that I found fascinating. Think about that: if you drop this book towards the floor, the Earth does respond by moving upwards (in the tiniest imaginable way, but it does move!).

My evening routine after primary school involved rushing back for some food prepared by my mum or grandmother. Generally speaking, we had a system called '6-8' each evening, where our dad expected us to be in our rooms, doing some reading or homework or something constructive before dinner. But before that it was fair game to do what we wanted! So, my typical weeknight would mean I watched:

5:00–5:10 *Newsround* (news but with a child-friendly angle)
5:10–5:35 *Blue Peter* (one of the longest-running children's TV shows in the world)
5:35–6:00 *Neighbours* (an Australian soap opera set in a fictional suburb of Melbourne, Victoria)

Now this is where I should have been back in my room, but sometimes my dad would stay late at work, and if he did, this was an opportunity for chaos! So the adapted evening schedule would be something like the following:

6:00–6:25 *The Simpsons*
6:25–7:15 *Star Trek: The Next Generation*

There are many fans of space who develop their love through watching or reading science fiction. But for me, that interest was already piqued through my primary years reading Russian-based books on the universe. So seeing *Star Trek*'s *Enterprise* Captain Jean-Luc Picard (brought to life by the commanding acting of Patrick Stewart) actually discuss the galaxies, stars systems and space phenomena that I had read about in books was jaw-dropping. This love for space was compounded when I used to collect the weekly *Star Trek Fact Files* magazine from WHSmith. There was a regular section on the numbers behind *Star Trek*: the speeds of starships, the engineering behind the ships and the physics behind how a 'beam me up Scotty' transporter might work. It was this that continued to keep me hooked on *Star Trek*.

OK, *Star Trek* allowed me to further fuel my love for numbers. But *Neighbours* probably didn't! At one stage in 1990, *Neighbours* was pulling in 21 million viewers in the UK, even going out twice on the same day (once at lunchtime and then again in the evening). I watched

Neighbours religiously from about the age of 6 to 19, so let's do some back-of-the-envelope calculations – estimating is always a useful life skill.

Including getting settled down before the programme and a post-show debrief with family or friends, let's say 30 minutes a day was allocated to *Neighbours*. So 2 hours 30 minutes a week (it wasn't shown at weekends). *Neighbours* would be on for most of the year – about 48 weeks. So $2.5 \times 48 = 120$ hours a year. With the 13 years of watching the programme, 13 years $\times$ 120 hours gives us 1,560 hours!

In Malcolm Gladwell's book *Outliers*, he claims that it takes at least 10,000 hours to became an expert in any subject. I had reached 15% of this target on an Australian soap opera alone. Who knows, had I devoted this time to juggling or learning languages, I could now be a world leading circus act or a diplomat at the UN. So any kids reading this, the moral of the story is to perhaps skip on the soap opera after school!

Programmes such as *Star Trek* may belong in the science fiction camp, but they serve a purpose beyond entertainment. First, they can inspire technologies of the future. We have seen electronic tablets, mobile phones, universal translators (Google Translate anyone?), and maybe a holodeck could be feasible with the invention of virtual reality goggles. Even the way people deal with each other points to a utopian society. In *Star Trek*, humanity seems to have overcome its continual aggravations with each other, and a universal harmony exists between all people. I think there is an element of the TV/film industry that tries to break barriers in our minds before it can happen in real life.

Today's science fiction is tomorrow's science fact. What is Hollywood for? To tease us with the rich and glamorous lifestyles of the actors and actresses that most mortals can only dream of? To entertain us? To make us think about what might be possible? For the 2014 epic sci-fi film *Interstellar*, director Christopher Nolan recruited theoretical physicist Kip Thorne (who ended up being awarded the 2017 Nobel Prize) to help him to make the science as accurate as possible. Films can inspire the next generation of mathematicians and scientists.

Science fiction can even make us ponder about how we would communicate with other species not from this planet. The 2016 sci-fi drama *Arrival* follows a linguist who learns how to communicate with aliens. But can mathematics offer us any insight into whether alien life might exist?

Dr Frank Drake, an American astrophysicist, founded SETI, an organisation dedicated to the Search for Extra Terrestrial Intelligence. Interestingly, right at the end of the twentieth century, SETI released a computer programme called SETI@home. This was designed to use the spare capacity in personal computers when they lie idle to do complex calculations related to SETI's search for life on other planets. As you might have guessed, our home was a proud participant, joining 290,000 other computers worldwide in the search for other life!

Recently, I have once again joined SETI@home to allow unused bandwidth in our internet broadband to support the listening for narrow-bandwidth radio signals from space. These signals are not known to occur naturally, and thus a detection on my laptop would provide evidence of extra-terrestrial technology (as yet, I have not found any aliens!).

In 1961, Drake came up with an equation to estimate how many detectable alien civilisations might exist in our galaxy, the Milky Way. It is not a rigorous equation, such as a Pythagoras theorem to work out the lengths of sides in a right-angled triangle, but it offers a wide range of possibilities.

His equation can also be seen as a back-of-the-envelope calculation, but the technical name for this is a Fermi estimation. Named after the 1938 Nobel Prize-winner in Physics, Enrico Fermi, this is an estimation technique that can allow us to make decent approximate calculations with little or no actual data. The most famous example is when Fermi estimated the strength of an atomic bomb test using the distance travelled by pieces of paper he dropped from his hand during this test.

The classic application of Fermi estimation is to estimate how many piano tuners there are in Chicago. Let's work through his logic.

1. Chicago's population is about 3 million people.
2. Assume an average family has 4 people so the number of families is 750,000.
3. Let's say 1 in 5 families owns a piano, so there will be 150,000 pianos in Chicago.

4. Let's assume that the average piano tuner services pianos 4 days a week (out of a 5-day working week) and has 2 weeks' holidays (Americans work harder than us Brits!).

So in a year (52 weeks), a piano tuner would service 1,000 pianos. So there must be 150 piano tuners as $150,000/(4 \times 5 \times 50) = 150$.

I've used this technique to get my class students to estimate how many bowls of cereal are eaten in the UK every year or the number of kilometres walked by people in Europe every day. If you make sensible assumptions, you can string them together to make some quite amazing predictions!

Now, let's look at Drake's equation to calculate an estimate for the number of civilisations in our galaxy that are detectable.

$$N = R^* \times f_p \times n_e \times f_l \times f_i \times f_c \times L$$

N = the number of civilisations in our galaxy that are detectable

R^* = the rate of star formation for the development of intelligent life

F_p = fraction of those stars with planets

N_e = number of planets, per solar system, with an environment suitable for life

F_l = fraction of suitable planets where life develops

F_i = fraction of life-bearing planets on which intelligent life emerges

F_c = fraction of civilisations that have developed communications (technology that releases detectable signs of existence into space)
L = the length of time over which such civilisations release detectable signals into space

As you read through this list, you may well be thinking that it's pretty hard to know the answers to plenty of those variables! The reality is that Drake wrote this equation to stimulate scientific dialogue as opposed to give us an exact answer. Nevertheless, here are the estimates that Drake proposed:

R^* = 1 star formed per year on average over the life of our galaxy
F_p = 0.2 to 0.5 (a fifth to a half of stars have planets)
N_e = 1 to 5
F_l = 1 (100% of these planets will develop life)
F_i = 1 (100% of these develop intelligent life)
F_c = 0.1 to 0.2 (10–20% of these can communicate)
L = 1,000 to 1,000,000,000 years

Using the minimum values, we get N as 20. Using the maximum values, we get 50 million. On balance, Drake and his colleagues came up with the estimate of 1,000 to 100 million civilisations in the Milky Way. Of course any single estimate is easily disputed but the Drake equation certainly got people talking at home.

It seems to me that if intelligent aliens do exist (so we'll rule out microbial life or extra-terrestrial slug-like creatures), they almost certainly wouldn't share our visions of reality. They are unlikely to appreciate the beauty of a painting by Leonardo da Vinci, a piano concerto by Beethoven, the sounds of a didgeridoo, a play by Shakespeare or even the pleasure of seeing a beautifully constructed set piece goal by Gareth Southgate's England football team. However, it seems much more likely they would understand the same maths as us.

Accounting for different number-based systems, an alien life form would still have the same prime numbers as us: 2, 3, 5, 7, 11 and so on. Maths might well be a more universal language of communication and it avoids the very human subjectivity of what makes good art or literature. Mathematical truths should be universal. So, in the film *Arrival* the linguists were the first to communicate with the aliens, but would a more appropriate person have been a mathematician?

But of course it's impossible to really know if that's the case. We assume counting is a natural phenomenon, because we can count the cows in our herd, the days between lunar cycles or the number of people in our tribe. However, if another species didn't have the same concrete experiences as us, such as living on a large gaseous planet, numbers might be a more fluid concept. So maybe mathematics won't turn out to be as universal as I think. But I do think it might be a better place to start a dialogue with our new alien overlords, rather than showing them a Picasso painting or playing a Mozart string quartet.

It increasingly seems that planet Earth may not be able to sustain human life forever. The final reason to leave may come through external threats (such as an asteroid strike) or internal ones (human impact on the world such as climate change). Even Stephen Hawking made the extraordinary claim that the human race only has 100 years before we need to colonise another planet. If as a species we are to start developing the futuristic technologies we need, maths needs to be at the heart of it. Elon Musk, the billionaire CEO of SpaceX, is determined for humans to prioritise the colonisation of Mars and has even suggested that 'up and down flights' with rockets to Mars should start in the first half of 2019, and that perhaps a million people could live on Mars by 2070.

The power of mathematics is ably demonstrated in the Drake equation. This allows us to estimate the number of intelligent civilisations in our galaxy. While looking out to space can seem like a deviation from our day-to-day lives, it does permit us the chance to use mathematics to think about what might be out there.

Chapter 5 Puzzle

Astronaut Training for *The Grand Tour* Crew

The Grand Tour cast of Jeremy Clarkson, Richard Hammond and James May have been enrolled in a slightly different adventure to the ones we are used to seeing on the road. They are in training to join an Elon Musk SpaceX visit to outer space.

As part of their astronaut training, they visit the following countries in this specific order with all 3 crew fitted snugly in a Mini car to recreate the cramped space shuttle conditions. Which country in Europe will they visit to finish off their training?

1. Greece
2. South Korea
3. Iran
4. Morocco
5. Canada
6. India
7. Paraguay
8. Mongolia

CHAPTER 6

'Maths Is Maths, and Art Is Art, and Never the Twain Shall Meet'

Pattern Recognition in Maths and Life

Picture this scenario: in 1991, my elder brother and I were voracious readers of fiction, particularly Roald Dahl (the J.K. Rowling of his day). Quickly establishing Dahl as our favourite author, we had pretty much every single Dahl book in the house (some on loan from East Ham library, others purchased from WHSmith and some bought at bargain rates from our local charity shop). Dahl's stories are distinctive for their wicked sense of humour and the inimitable illustrations of Quentin Blake. So the only logical thing for a 7-year-old and his 10-year-old brother to do was to re-draw every single Blake illustration from the books. Again, again, and again. In fact, we copied so many of Blake's pictures that we would fill entire cardboard boxes with drawings. Our father collected discarded paper from his office to satisfy our monstrous appetites.

But that is not the end to the story. We then packaged up many of our hundreds of drawings and posted them off to Quentin Blake himself, not expecting much. Cue whoops and cheers at home when a few months later, we received a personalised hand-written letter of thanks from the artist himself, with illustrations in that uniquely Blake style. Over the course of the next year or so, we proceeded to shuttle across more drawings and Blake replied with signed books and more letters and illustrations.

What does art have to do with maths or numbers? I often think of Rudyard Kipling's 1889 poem 'The Ballad of East and West', where 'East is East, and West is West, and never the twain shall meet', in reference to how many people think about art and maths. But that was never the case for me. When I wasn't playing about with numbers as a child (calculating the speeds of my favourite sprinters or comparing the permutations of different end-of-season football league scenarios), I could often be found sitting in my room attempting to do watercolour imitations of Rembrandt's self-portraits or crayon sketches of a futuristic *Knight Rider* car.

When I was only 7 years old, my dad bought me 2 books aimed at introducing young readers to the life, art and works of Rembrandt van Rijn and Pablo Picasso – part of the Scholastic children's series of *Getting to Know the World's Greatest Artists*. (Yes, 7 does appear to be a moderately tender age to be introduced to such renowned artists, but I was also presented with Dennis the Menace comics around the same time. Needless to say, I was equally gripped with Dennis.) I was entranced by the contrast between Rembrandt and Picasso: one utterly precise and realistic in his depictions, and the other imagining his own versions of reality.

As a child, I drew comfort from the absolute correctness of maths that one encounters at that age. Mathematical calculations such as addition and multiplication always yield the same result – no matter how much of a bad day you're having! Yet, Rembrandt and Picasso were admired as two of the greatest artists ever and I found it bemusing they could have such a different approach to art. (As we'll discover later in this book I also became entranced by a third book in this series, on Leonardo da Vinci and the sheer breadth of his accomplishments as an artist, architect, inventor, even scientist, a true Italian Renaissance polymath, a man who was able to straddle genius in multiple fields.)

In summer 2017, my *University Challenge* pal Eric Monkman and I recorded a programme for BBC Radio 4 on Polymaths. As part of this programme, we met with Professor Stefan Collini, a professor of English Literature and Intellectual History at Cambridge University. We discussed 'The Two Cultures', the famous lecture delivered in 1959 by British scientist and novelist C.P. Snow. Snow felt that western society was split into 2 cultures, the sciences and humanities, and that this was a significant stumbling block to solving world problems. Any such divide in disciplines never occurred to me as a child – I never once considered the complexities of being comfortable both with the certainty of numbers and the ambiguity of art. Sometimes as adults, we pigeonhole people as arty or scientific as if they are mutually exclusive frameworks. But as I have grown older, I have come to realise that the worlds of art and mathematics have both informed each other a great deal, even if some in those fields may not be immediately aware of it.

One key skill in mathematics is spotting patterns and then being able to recognise the same pattern occurring in other situations. And in many forms of art, patterns are layered and repeated to produce interesting new effects. As a child, I was fascinated with the idea of repeated patterns – whether it was on wallpaper, wrapping paper or even a duvet cover. There's a particular postcard that sat on one of our mantelpieces, perhaps purchased from one of our numerous visits to the local charity shop on the way back from our Saturday pilgrimage to the library. This postcard drew second and third glances from me for years. It was a black and white pencil sketch of an impossible and dreamlike world that seemed to defy the basic laws of Newtonian physics and gravity. The picture reminded me of a scene from the upside-down staircases that the Goblin King (played so hauntingly by the late David Bowie) walked across in the 1986 dark fantasy film *Labyrinth*. It wasn't until secondary school, at the age of 11, that I came to more formally understand the significance of this work and its creator: M.C. Escher.

Maurits Cornelis Escher was a twentieth-century Dutch artist who made mathematically inspired creations in the media of woodcuts and lithographs. Escher explored the concepts of impossible objects, infinity, reflection, perspective and tessellations among others. I found him to be the most mathematically fascinating artist. Trying to imitate his work was challenging, requiring discipline and focus, unlike the free strokes I employed when copying Quentin Blake's illustrations.

Two works in particular demonstrate how Escher intertwined mathematics and art in this magnetic form, and they're easy to find online. The *House of Stairs* appears impossible to construct in real life, as it had 2 vanishing points and made parallel lines run towards and around them in a curved manner. The top of the lithograph print is a repetition of the bottom, so the scene could carry on forever, much like wallpaper or mathematical infinity. A second work, Escher's *Circle Limit IV*, had angels and demons form a tessellating pattern. The spaces between one cleverly form the shapes of another. This is similar to his *Circle Limit III*, which our class teacher when I was age 7 curiously used in lesson demonstrations without perhaps realising that it was an inadvertent early introduction to hyperbolic geometry. Here, the pattern of fish shrinks towards the edge, seemingly continuing to infinity. This woodcut represented an impossible 2-dimensional surface known as a hyperbolic plane. Again, with my interest in maths, I was immediately drawn to these images, which seem to represent concepts that I only understood abstractly.

Perhaps I was hardwired to appreciate this kind of aesthetic.

If you take a close look at a daisy flower, with its fancy Latin name of *Bellis perennis*, you will see how the petals spiral. The number of petals in each row is the sum of the preceding 2. It's called a Fibonacci sequence. This was discovered in 1202 by the Italian Leonardo Fibonacci. While contemplating the mathematics of happily breeding rabbits, he observed a sequence formed through constantly adding together the previous 2 terms: 1, 1, 2, 3, 5, 8, 13, 21, 34, 55 and so on. As the numbers get larger and larger, the ratio between 2 numbers next to each other tends to 1.62. This is a special number that since the beginning of time has hidden in plain sight around us, embedded in nature, and some may even say is fundamental to life itself – the ratio of the major groove to the minor groove of a DNA spiral is roughly 21:13 (and 21 divided by 13 is 1.62).

The Greeks knew of this number and called it the Golden Ratio and we assign it the letter phi.

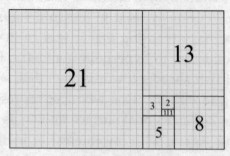

Wikimedia Commons

The Golden Ratio can be seen in a myriad of different places in the world around us. If you can imagine how a nautilus shell grows, the way in which a violent hurricane builds or the movement of a distant galaxy, they all curl according to this same Golden Ratio. Some believe that the Golden Ratio even shapes our bodies, as it describes the ratio of our overall height divided by the height to our navel, the ratios of the lengths of bones in our fingers, and the ratio of the length from our elbow to the length of our hand.

Upon first encountering the Golden Ratio, it did truly seem divine to me. Finding this ratio crop up everywhere from the *Mona Lisa* to the front of the Parthenon to the shape of the galaxy seemed too coincidental. The cynic in me thinks sometimes maybe it *is* all just a coincidence. As a *Doctor Who* fan, it is an apt moment to refer to his quote from the 1996 film where he expressed, 'I love humans. Always seeing patterns in things that aren't there.' However, the romantic in me (both mathematician and artist) thinks that the frequent appearance of the Golden Ratio may be one of the great mysteries of the world, and we'll never quite know what it means.

Some of the most intractable problems in mathematics are known as the Millennium Prize Problems. Set up by the Clay Mathematics Institute in America, these 7 problems have 'resisted solution over the years'. Currently, only 1 problem has been solved and the other 6 still remain open. For a solution to be accepted for a prize, a solution needs to have general acceptance in the mathematics community for 2 years after publication. Complex mathematical proofs may only be understood by a handful of experts in that particular field. Gaining peer approval of the proof is the rubber-stamp a mathematician needs for their proof to gain acceptance and recognition (and perhaps public adoration!).

I wonder whether a similar authentication period should be used in the world of art when 'lost masters' are rediscovered. Every so often we hear of a Picasso or a Degas or a Constable being unearthed in a car boot sale or from the attic of an elderly grandmother. How does one move from the status of claiming something as the work of a great artist to it becoming widely accepted as genuine? This was something I considered back in 2011.

For me, the blockbuster art exhibition in London during 2011 was 'Leonardo da Vinci: Painter at the Court of Milan', which the National Gallery claimed to be the most complete display of Leonardo's rare surviving paintings ever held. Having been a fan of da Vinci since receiving another art-based history book as a 7-year-old, I remember feeling the hairs on my arms standing on end, quite literally, when I first heard about this exhibition. In December that year, I went to see it. Like a mathematician revisiting a beautiful proof he had seen before, I then could not resist seeing this exhibition for a second time.

The primary draw of the exhibition was the unique opportunity to see both versions of da Vinci's *The*

Virgin of the Rocks staring opposite each other in the same room, perhaps for the first time ever. Maybe even da Vinci himself never had this pleasure. One version resides permanently in the National Gallery but the other was brought over from the Louvre in Paris. The paintings look identical to the casual passer-by, but a second glance would confirm significant differences. As a maths teacher, I have taught students to understand when shapes, particularly triangles, are identical or mathematically congruent, where all the corresponding sides and angles are equal. These 2 paintings would perhaps be called mathematically similar, close but no cigar.

Near *The Virgin of the Rocks* was a painting of then-disputed provenance, the *Salvator Mundi* (the Saviour of the World). It was long thought to be a copy of a lost original but then it was restored, re-evaluated and proclaimed to the world as an original of da Vinci. Despite disputes from several scholars, the painting smashed the highest price paid at auction for a painting in November 2017, with an Abu Dhabi group spending an eye-watering $450 million to purchase it.

Assessing the authenticity of a piece of artwork can involve the expertise of art historians and curators. However, in certain cases, mathematics can play a role. There are some works of art – perhaps da Vinci's *Mona Lisa* or Michelangelo's towering statue of David – that are fairly universally recognised as exquisite constructions, that required prodigous craft and execution by the creator.

CHAPTER 6

'Random splats on a canvas that anyone could have created.' These were my, perhaps naïve, first thoughts on seeing the paintings of the influential and provocative American artist, Jackson Pollock. He was renowned for having pioneered action painting, a technique that saw him drip paint on canvases lying flat on a studio floor. In 2006, a Pollock work (*No. 5, 1948*) sold for $140 million, at the time the highest price for a painting. To the untrained observer, it seems fair comment to think, 'What's the big deal? I could do that myself.' Compared to the fine dexterity required by da Vinci, this seemed like child's play. At one stage in school, I tried to be good to my word and 'do that myself', but my attempts at throwing paint with my brush onto paper failed to recreate Pollock's art in the way that I had hoped!

Mathematics actually illuminates a greater depth of thought to the work of 'Jack the Dripper', a moniker given to Pollock due to his technique. Pollock captured some aesthetic dimensions, a reasoning of human perception, which was perhaps beyond the scope of some of his critics. So here comes mathematics to the rescue. In 1999, physicist and art historian Richard Taylor from the University of Oregon led a crack team of mathematicians to analyse the paintings, and uncovered that the logic of Pollock lay not in art, but mathematics. Specifically, chaos theory and its progeny, fractal geometry, came to the fore. Fractals can seem haphazard at first, but each one is a single geometric pattern that is repeated thousands of times at differing levels of magnification – a bit like an ornately crafted wooden Russian doll, one nested inside another.

Taking a more detailed look at Pollock's work, magnified sections appear to be very similar to the full-scale versions and show the characteristic of fractals, that of infinite complexity. Of course, when we keep zooming in on his work perhaps a thousand-fold, as we could on an iPad, then we would just see individual blobs of paint. In mathematics, we can use fractal dimensions to describe these patterns and this can be applied to Pollock's technique. The drips, blobs and splotches on Pollock's works seemingly create repeating patterns at different scales.

Therefore it is maths we need, both to understand Pollock's work and to test whether a disputed work is a genuine Pollock rather than just a child splashing paint chaotically. We all know the dimensions that a line, a square and a cube each has: 1, 2 and 3 respectively. A line has distance, a square has area and a cube has volume, as my 12-year-old students will testify from my lessons on dimensions. But what happens when we think about the dimensions of the inside of the lungs or the brain: how can we measure their surface area? What about a piece of Romanesco broccoli? This vegetable shows a fractal-esque nature. The fractals show self-similarity or a comparable structure regardless of the scale. So if we imagine a small piece of this variety of broccoli, when we view it up close, it will look the same as a larger chunk. Fractal dimensions allow us to measure the complexity of an object. A cauliflower comes with a dimension of 2.3, broccoli at 2.7 and the surface of the human lung just beneath 3. (Of course these are not true fractals because at high levels of magnification, the object will lose its self-similar shape. At this stage, it will reveal plain old molecules.)

Pollock's early paintings had a fractal dimension of 1.45, about the value for the fjords in Norway, but his last works had a fractal dimension of around 1.7. In order to replicate Pollock's works, you would need to consider his height, physiology and muscular build-up, and spend years learning about refining patterns. There are estimated to be about 400 fake Pollock paintings in circulation. Taylor has developed a mathematical algorithm that can be used to verify the authenticity of Pollock's artworks. Of course, there has been some resentment in the art community, saying that you can't use maths and numbers to quantify the emotional state of humans in creating art. However, Taylor has made clear that while computers can look at the patterns of a Pollock and establish credibility, they cannot tell you whether to prefer one artist over another (well, not yet anyway). It's still your call whether to have a replica Constable landscape or a Picasso cubist picture up on your wall.

Like mathematics, getting good at art is something that takes hard work, diligence and persistence. Even though everyone starts with a particular set of skills, you can make great strides with a lot of practice and dedication. I sometimes wish that my school art lessons had not stopped after Year 9 when I was aged 14. At the time I suspect I thought pragmatically that I could only select 2 of 4 subjects for my GCSE options at 16. If I could go back in time, I would tell younger Bobby aged 14, go and pick art! Perhaps one day in retirement, you'll find me picking up my pencils, crayons and pastels once again, and trying to recreate the carefree art of my youth.

As a mathematician trying to understand the world through a framework of numbers and patterns, I think art is another way to help see the world around us with fresh eyes. For one person, a mathematical formula may describe the trajectory of England footballer Harry Kane's header arching into the back of the net, and for another, a drawing captures that same moment of joy as a ball curves towards the goal. But as for me, I love both!

Chapter 6 Puzzle

Damien Hirst's Mathematical Spot Paintings

Let's engage your mind for numbers. Some of the most recognised works by modern artist Damien Hirst are his spot paintings. He has been inspired by mathematics to create a new pattern of dots on his canvas. This is being hailed as the greatest modern art since Picasso's cubist revolution!

Hirst paints dots in neatly laid out columns. In the first column he paints 3 dots, in the second 4 dots, in the third 6 dots. The number of dots actually forms the following sequence: 3, 4, 6, 8, 12, 14, 18, 20. How many dots does Damien Hirst paint in the next column?

HINT: You don't need to be in your mental prime to solve this!

CHAPTER 7

Ready, Steady, Cook

A Splash of Numbers in the Kitchen

Cooking is an art form, not a precise science. We eat for pleasure, not just to fuel our bodies, and many people cook for pleasure as well, not just as a way to make food appear on the table. But some people take it much more scientifically. The branch of food science called molecular gastronomy seeks to understand the physical and chemical transformations of ingredients that happen in the cooking process. There are 3 components of this field: social, artistic and technical. We have to accept that while there are scientific factors at play when you cook a fried egg (viscosity, surface tension, how to introduce air), there is also a human element of creativity and individuality that cooking requires – something I haven't quite mastered yet. However, for us to reach these levels, we need to master the basic numerical skills for the kitchen whether consciously or subconsciously. Show me a cook that is incompetent at numbers and I will show you someone who is an incompetent cook. They may not realise it, but every cook is a mathematician too.

During my primary school science lessons, I encountered Mrs Gren. She isn't a teacher, or a lovely lollipop lady who helps children across the road, but an acronym to help remember the characteristics that differentiate living things from material objects. The acronym MRS GREN stands for movement, respiration, sensitivity, growth, reproduction, excretion and nutrition. It's the last of those terms that I want to focus on in this chapter.

In order for organisms to survive, we require food for energy and nutrients. Ultimately, our mitochondria cells convert oxygen and nutrients into ATP (adenosine triphosphate), which is the chemical energy 'currency' that powers the metabolic activities of the cell. So without nutrition, we would quickly wither and waste away. Without sufficient nutrition, our bodies would fade away and our minds, able to perceive and conceive maths, would cease to exist.

If you have a flick through the TV guides on any evening in the UK, you'll notice how popular cookery shows have become. Whether it's *MasterChef* or *Bake Off*, Gordon Ramsay or Nigella, there's something for everyone. Look in any bookshop, and you'll find culinary names peppering the bestseller lists. From TV screen to Instagram photo feeds, it seems that, as a nation, the UK loves cooking.

The most basic nutritional mathematics is that if we want to sustain ourselves, our calorific intake via grub and beverages must equal calorific expenditure through our exertions (mental and physical). Of course, those looking to lose weight need expenditure to exceed intake and vice versa for those looking to put on weight: intake should exceed expenditure.

My relationship with cooking is a mixed one. There used to a be a cooking TV show in the UK called *Can't Cook, Won't Cook* where 2 contestants (chefs for the duration of the show) were nominated by friends or family to cook under the supervision of a top-class chef. As hinted at by the title, the 'can't' cook people claimed their abilities were limited to burning toast, and the 'won't' just simply never made time for it or refused to try. I would call myself as a hybrid version of these 2 – on the whole I don't make time for cooking, but then again, I did give my younger brother food poisoning by cooking that most complex of dishes, a fried egg!

My mum tends to cook south Indian dishes from Kerala with recipes that have been handed down through the generations. My father often eats an unusual combination of Indian fusion mixed with whatever is in our fridge/freezer – each meal tends to be one of a kind! I sadly have not learned Indian cooking (yet) from my parents sufficiently to be an independent Indian cooking wizard, but I am a helpful hand in the kitchen. I can claim that some of my early love for numbers came from being a great sous chef and helping out.

From a young age, helping out in the kitchen taught me about quantities, about proportions and categories, as my parents asked me to add more coriander to the salad mixture, or make the dough balls for the chapatis bigger or smaller. Sorting items into different groups was my earliest inadvertent exposure to set theory (remember Venn diagrams from school?) – sorting ingredients into matching piles, the red peppers in one corner of the chopping board and the green peppers on the other side. Breaking eggs was not my forte as a child, and it is still something I cannot do without the prospect of tiny fragments of shell scattering in the bowl. However, Indian spicy omelettes are very common and it was easy to develop counting skills as my parents broke the eggs in the bowls: 1 egg, 2 eggs, 3 eggs and so on. (We did have a set of scales, though in an Indian family kitchen, much of it is done by eye rather than measurements.)

My development in the kitchen did not progress much further till I joined my secondary school, St Bonaventure's (known to everyone as St Bon's). Every academic year between Year 7 and Year 9, all students would have a half term's worth of Food Technology lessons. Mrs Savile was our Food Tech teacher, a small yet perfectly poised lady who gave precise instructions about what to do and when. After her demonstrations on how to make a rock cake (a fruit bun with a sturdy texture and craggy exterior), we were set loose with our instruction sheet. I remember meticulously following the instructions. Despite not using the set of scales at home, I found refuge in the precision of the numbers in these lessons. Whereas some of my friends were using judgement and intuition, I found comfort in measuring exactly 200 grams of self-raising flour and 100 grams of butter, no more no less according to the scales. This did result in me being sluggish in pace compared to many of my peers during Food Tech lessons.

At the end of every Food Tech lesson, we followed the same washing-up routine. Washing bowl out, a dollop of Fairy Liquid and half a tub full of warm water. Then allow the dishes to soak for a very short period before using a scrubber to remove tough stains and finally rinse underneath the sink tap. I found joy in this routine – the order and precision of it, almost like the set procedure of a mathematical calculation bringing everything together. The lessons did spark a renewed interest in assisting my parents in the kitchen, though I found that it was the manual dish-cleaning that particularly drew my attention.

Years later, my most serious adventure in the kitchen was on a 2-week course. I was in between jobs after my time as a trader in investment banking and the start of my graduate trainee accountancy programme at PwC. It was organised by the Young Foundation, a progressive think tank that ran projects for young people. This particular one was called 'Faking It', giving young people in the area the opportunity to 'fake it' as a chef. After a 2-week intensive programme, we cooked in an actual professional kitchen in a restaurant in Hoxton for a group of visitors. Again, flair can often follow a careful attention to the numbers, but I found that I was still an excessively cautious slave to the measurements. I wondered whether it was possible to marry my fastidious attention to the recipe calculations with an understanding of how to create food that really did delight the taste buds?

Step forward an inspiration on my TV screen, mathematician, pianist and food enthusiast Dr Eugenia Cheng. For those who haven't seen her, her enthusiasm for maths is something to behold! I had the pleasure of introducing her to an audience at the Royal Institution in 2018. She was on the *Late Show with Stephen Colbert* in the United States in November 2015, talking about how to make numbers delicious and introducing her most recent book, *How to Bake Pi*. Colbert's show attracts some big-hitters, including actor George Clooney, former Secretary of State Joe Biden and President Obama. So for a mathematician to get invited to talk about baking and numbers is pretty special.

On the show, Dr Cheng was helping Colbert to make mille-feuille, a sweet French pastry consisting of 3 layers of puff pastry alternating with a double layer of pastry cream, which literally translates as 'thousand-leaf'. Sounds delicious or 'miam miam' as the French might say! The power of maths actually shows why the name of this pastry reveals some basic maths. If you take some pastry and fold it on top of itself, we have 2 layers of pastry. If we fold this into 3, we now have 6 layers (2×3). Roll out the pastry again and then fold into 2 once more to give us 18 layers (3×6). If we repeat this again, again, again and again, we get $18 \times 3 \times 3 \times 3 \times 3$ which is 1,458. So just with some small numbers – 2 and 3 – we are able to develop a pastry with 1,458 layers (hence the name of a thousand-leaf). This is the power of exponentials in maths: $2 \times 3 \times 3 \times 3 \times 3 \times 3 \times 3$, which is 2×3 to the power of 6. Quickly the number of layers increase from 2 to 6 to 18 to 54 to 162 to 486 and finally to 1,458.

On the show, when Dr Cheng was calculating this with Colbert, she wasn't able to work this out in her head, to which she responded beautifully, 'I'm a mathematician, not a calculator.' This gets right to the heart of the distinction between numeracy and mathematics. Numeracy is about our competency with our arithmetic skills – addition, subtraction, multiplication and division – the day-to-day skills we need to function. Maths is more than that. Maths is about being able to spot patterns, whether in the real world or abstractly. Calculators have long been quicker and more powerful at numeracy than the best humans. But mathematics requires true thinking and something that computers can't do without our help.

So while mathematics and numeracy aren't quite the same, numeracy is a subset of the wider group of mathematics. The vast majority of us will cook and many of us will bake, and therefore use numbers on a day-to-day basis. Due to the prevalence of cooking, you could say that we are a nation of mathematicians who do not quite think of themselves in that way. Certain parts of mathematics are about following an algorithm, a set of instructions in a particular order, and following a recipe is no different. Creating your own recipes is like creating your own maths. You use principles that you have assimilated to come up with fresh or new interpretations.

Unless you are solely making cold salads, you will need to use the oven for your master chef impressions. When using an oven, conversions are critical. One of my first attempts to bake was as a child aged about 7. Despite paying exacting attention to the ingredients being added, my concoction came out of the oven so solid that it could have been an offensive weapon! I made the classic error of misreading the temperature – I think I mistakenly read the higher Fahrenheit amount instead of the Celsius that our cooker was scaled in, and the resulting temperature was hot enough to pulverise our cake.

The lesson is ensuring that you get your conversions spot on. While there is manoeuvrability in the kitchen for an extra dash of salt or an additional splash of oil on the pan, using the wrong temperature scale will be catastrophic for your dish. Sadly, there is no universal convention on what measure of temperature to use. In the UK, cookers tend be in Celsius or the 'gas mark' system as well. Celsius tends to be used in pretty much all the countries in the world. The exception is the use of Fahrenheit for our cousins across the pond in the United States and curiously also in Myanmar and Liberia. A pub quiz fact for you is that Celsius and Fahrenheit are equivalent at −40 degrees, in case you ever wanted to know!

Nowadays with the use of Google, we can work out the conversions between the 2 in a flash, but it is always useful to understand how you would do it manually – especially in those moments our Wi-Fi or 3G signal fails us. To convert from Celsius to Fahrenheit, we use the formula: $F = 9/5\,C + 32$

If we want to covert 100° C into F, we substitute the 100 into the formula.

$$F = 9/5 (100) + 32$$
$$F = 180 + 32$$
$$F = 212$$

To go the other way around, from Fahrenheit to Celsius requires some reverse engineering, or as mathematicians say, rearranging the equation. Currently F is the subject, but we want to make C the subject: $C = 5/9 (F-32)$

Let's have a go at converting the 212°F back into C.

$$C = 5/9 (212 - 32)$$
$$C = 5/9 (180)$$
$$C = 5/9 \times 180$$
$$C = 100$$

Of course, most cookers have an analogue dial, which only has every 25°C or so marked out. So this requires sensible estimation on behalf of the cook.

Weight is an important consideration in cooking. At home, Christmas turkey tends to be the main festive source of protein on 25 December (with some left over for the next day). Cooking a 7-pound (about 3.2 kilograms) turkey requires some basic proportion understanding. If the turkey is required to thaw in a refrigerator for

24 full hours for every 5 pounds, you need to plan in advance and get the turkey out of the freezer in sufficient time. So let's see the proportions in use: $5/24 = 7/x$ [this is pounds/hours]

This basically says that 5 pounds is to 24 hours as 7 pounds is to x amount of hours. This requires a bit of cross multiplication of fractions. So $24 \times 7/5$ gives us 33.6 hours. With this, we can't afford to make an error such as doing $24 \times 5/7 = 17.1$ hours. Using this, we would take the turkey out to defrost far too late, unless you wanted it ready for breakfast on Boxing Day instead! When I teach students in school, often the most challenging part can be helping them to develop a number sense, where we can look at our answer and think, 'Hey, this looks wrong.'

If we are chained to our calculator, mindlessly entering numbers into a formula without understanding, then we are unlikely to spot glaring errors. With the turkey example, hopefully most adults would think that if they got 17 hours as their answer for cooking a 7-pound turkey, this seems out of whack. As it takes 24 hours for a 5-pound turkey, you need more for a heavier turkey. Hence the 17 hours must have involved a miscalculation.

Again this use of proportion is handy if we are looking at instructions about length of cooking time. The turkey instructions may say that it needs 25 minutes per pound, so we obviously need to work out how long it takes for our turkey. This is a more basic ratios question. Twenty-five minutes per pound for 7 pounds means multiplying $25 \times 7 = 175$. The turkey needs just shy of 3 hours, clocking in at 2 hours and 55 minutes. Though this obviously does not include the additional time for warming the oven.

Changing quantities is one of the most frequently used bits of mathematics in the kitchen. You may be baking a batch of cookies with a recipe from a Nadiya Hussain cookbook (she was the superstar winner of the *Bake Off* cooking series in 2015). If Nadiya's recipe is for 6 cookies, but you want to make say 18 cookies, then you have to multiply all ingredients by 3 to make your larger batch. This sounds glaringly obvious, but this is an application of maths. Things get more challenging if it involves fractions. If the recipe asks for 2/3 of a cup of milk, then you are going to have to multiply $3 \times 2/3$ to give you 6/3 which is 2 cups. Of course, do make sure you know what the size of the cup is in the first place!

Cooking and mathematics may not seem like instinctive bedfellows, yet cooking ultimately involves following sets of procedures allied to numbers within specific timeframes. A palate for cooking flavours and a mind for numbers combine together to make some of the most delicious recipes of our society.

Chapter 7 Puzzle

Ronald McDonald vs Usain Bolt: Chicken McNugget Challenge

I am a fan of McDonald's fast food and have witnessed Chicken McNugget Challenges. A brave eater will attempt to consume 100 McNuggets in one single sitting. The sprinting legend Usain Bolt gets into a challenge with McDonald's mascot, Ronald McDonald.

Ronald McDonald eats 1 McNugget on Monday, then doubles his intake to 2 McNuggets on Tuesday, and then doubles again to 4 McNuggets on Wednesday. He keeps doing this, doubling his intake each day, so after 3 days, Ronald eats $1 + 2 + 4$ McNuggets, a total of 7 McNuggets.

Usain Bolt starts off with 10 McNuggets on Monday, then 20 on Tuesday, then 30 on Wednesday and so on, increasing by 10 McNuggets each day. So after 3 days, Usain eats $10 + 20 + 30$ McNuggets, a total of 60 McNuggets.

After how many days does Ronald eat more McNuggets in total than Usain?

CHAPTER 8

'Thank You for the Music'

The Numbers that Power Our Listening

Students say Mr Seagull, you can call me Bobby,
You see maths for me, is more than just a hobby.
Two twos are four, and four twos are eight,
Starting with your tables will be just great.
Trigonometry is all about the angle,
Ratios sin cos but don't get in a tangle
Area of a circle, pi r squared,
Pi times 2 r, circumference if you cared.
Y equals mx plus c is a just straight line,
M the gradient, c the intercept be mine.
Numerator over denominator,
Get it right or Mr Seagull will you see later.

This was my attempt at a maths 'rap' (alongside Canadian rapper Drake's instrumental beats) in order to engage my school students in a maths lesson. In late 2017, one of my Year 11 pupils (15–16-year-olds preparing for GCSE public exams) was bobbing his head up and down as he was ploughing through some exercises on circles. As a teacher surveying the class, you develop a sixth sense for when a student might be distracted, and I thought this bobbing was a signal of disturbance. As it turned out, the student was on task but he said his head movement was because he was thinking of music bars and beats at the same time as working. He added, 'Sir, I've got a great one for you: "your name is Mr Bobby and maths is your main hobby".' My response, as a diligent teacher, was to politely tell him to get focused back on his work, but that I'd consider using that in a rap to help them learn some maths concepts.

I'm an east Londoner who has tried to remain in touch with contemporary London music, and my students know I'm a fan of grime, rap and hip hop. However, it was they who introduced me to the 'quick maths' of comedian turned rapper Big Shaq (real name, Michael Dapaah). In the early part of the academic year 2017–18, students would be rapping the opening lyrics to his song 'Man's Not Hot' with the words '2 plus 2 is 4, minus 1 that's 3, quick maths' echoing round the playground (and sometimes my maths class!). They could recite the rest of the song with ease, and yet these same students would struggle to recall basic mathematical operations, such as times tables, or key formulae such as the area of a triangle. This made me reflect that music has such a power, both to entertain us but also to resonate with us in a way that makes things memorable and stick them in our minds.

Apart from beginning with the same initial letter, is there a link between mathematics and music? Of course, when learning to play a musical instrument, you have to understand fractions and ratios to count beats and rhythms. The most obvious mathematical aspect to music involves the metre such as 3/4 time or 4/4 time where musicians will count beats to a bar. And our friends across the pond have made it much easier to understand how all this works – over here, musicians talk about 'crochets' and 'minims' for notes lasting 1 or 2 beats. Our American cousins are far more logical and call them 'quarter' and 'half notes'.

The most common metre is 4/4 time, where there are 4 quarter-note beats. So to count in 4/4 metre, we have to tap the beat, which is the equivalent of 1 quarter-note – such as in the song 'Old MacDonald Had a Farm'. The second most common metre is the 3/4, where the combination equals 3 quarter-note beats, as used in 'Here We Go Round the Mulberry Bush'. Mathematics can help learners understand the value of music notes, such as half notes and eighth notes, and their equivalent fractional size.

Gottfried Leibniz, the seventeenth-century German mathematician and rival of Sir Isaac Newton, wrote that 'music is the sensation of counting without being aware you were counting.' The fundamentals of mathematics start with numbers and, in an analogous manner, the grammar of music – rhythm and pitch – owes much to mathematics. We may hear 2 notes an octave apart and get the sensation of listening to the same note. This is due to the frequencies of the notes being precisely in the ratio of 1:2.

The jury's still out on whether being good at music helps you with maths, or whether being good at maths helps you with music. However, anecdotally, many of us have known someone at school who was able to play a couple of musical instruments, and showed high competence in maths.

Although I'm a growth mindset kind of person, who believes that virtually everyone can improve their competence in most fields with enough deliberate and targeted effort over a period of time, playing musical instruments is an area where I'll confess I've never had much natural talent! Though I love listening to all kinds of music, and find it really powerful and emotional, playing music has always been something I've struggled to do well.

A victory for my beloved West Ham? A slightly off-key rendition of 'I'm Forever Blowing Bubbles' will evoke feelings of euphoria in me. Switch on the news and see the tragedy of the 2017 Grenfell Tower fire unfold in London? The mournful notes of Samuel Barber's 'Adagio for Strings' leaves me quiet and reflective. Getting ready for a night out on the town? Some readers might disagree, but I find English pop sensation Dua Lipa's upbeat pop hits such as 'New Rules' are the perfect thing to get me appropriately hyped up. Music has the power to resonate with our soul.

I don't necessarily believe in having regrets. You make decisions, you abide by them and make the most of what you have going forward. But I do wish I'd made more of an effort to play musical instruments when I was younger. At age 5, my first exposure to playing an instrument was when my father bought a Casio keyboard. I learned various Disney songs (such as *Pinocchio*'s 'When You Wish Upon a Star' and *Snow White*'s 'Heigh-Ho') by following lights above the keys. The actual method of learning seeming quite algorithmic – little lights would flash above the key you were meant to press, and you would push down the keys and follow along. Eventually, if you were able to repeat this sequence of notes, a melody resembling the tune would be recognisable to those nearby. It did remind me of early times-table learning, perhaps not quite understanding what the relation between numbers meant, but memorising them nonetheless.

It was only at secondary school, aged 11–14, that I formally learned how to read music and started having weekly 30-minute piano lessons. By the third year, I had made slow and steady progress, but decided to give it up before my Grade 3 exams ahead of my final 2 years at school. While there is a spectrum of musical talent out there, where some people have an instinctive sound for notes and pitch, practice and effort can improve your individual standing. I improved my mathematics through effort but, for some reason, didn't quite apply myself to the piano.

So I gave up *playing* music but I still loved *listening* to it.

Growing up in the 1990s, and obsessed with numbers and lists in general, I would be glued to my TV screen on a Friday evening for *Top of the Pops*. This was a music chart show that would count down the top-selling music singles to the coveted number 1 slot. In the 1990s, as music was purchased on tapes, CDs or vinyl records, it was straightforward to count record sales and order rankings. As new music streaming services have become more and more widespread, music charts have had to include online streaming figures into their chart calculations.

For physical copies of singles, the top global positions are taken up by Elton John's 1997 'Candle in the Wind', which sold 33 million copies, and Bing Crosby's 1942 'White Christmas' sitting pretty at the top with 50 million sales. Due to the shift from physical sales to streaming, these are unlikely ever to be toppled.

So how are modern singles charts calculated? To stream a song, you log onto your platform such as Spotify and then hit play. Currently 150 individual streams equate to 1 sale in the chart compilation process. So if a track gets 150,000 streams across all the different platforms, this is counted as 1,000 sales by the charts (150,000 divided by 150). Of course, digital downloads and physical sales are added onto that total but these numbers are diminishing as 'streams' continue to grow. Prior to January 2017, the ratio was 100:1. The 'Official Charts Company' of the UK review this rate continually to reflect the rise of streaming. In 2014, the number of streams per week was 275 million, but by 2016 it had reached an incredible 990 million and this number continues to head on an upwards trajectory. Acknowledging the rise of music videos on platforms such as YouTube, from July 2018, 600 plays on advert-supported services also equate to 1 sale. For plays on paid-for services such as Spotify Premium videos, the ratio is 100 plays to 1 sale.

Album chart calculations are slightly different. One of my earliest physical album purchases was a CD of Britpop band Oasis with their 1994 *(What's the Story) Morning Glory*. Nowadays, every time an album track, such as 'Wonderwall', is streamed, it will count towards the album's sales records. As of 2015, 1,000 plays of an album would equate to 1 physical sale or download of an album. However, we know from our personal music listening experience that certain tracks get streamed many more times than other 'filler' tracks (compare 'Wonderwall' on Oasis's 1994 album to the filler song 'She's Electric').

Here too, the Official Charts Company constantly re-evaluates the arithmetic behind its calculations. To prevent smash hit songs like 'Wonderwall' skewing their numbers, they created a system of 'down-weighting' the top 2 songs per album. The current trend on streaming sites is for consumers to listen to their favourite songs again and again. So they work out the arithmetic average or mean of the number of streams for the songs (excluding the top 2) on the album. Then they move down the streaming number contribution of the top 2 songs to the average of the other 10. Finally they get a total number of streams and divide by 1,000 (before adding to any digital album downloads and physical sales). As with single streams ratios moving from 100:1 to 150:1, over time we can expect the 1,000 for albums to increase to 1,500:1 and 2,000:1.

However, the mathematicians behind the charts have had to respond to other distortions of streaming. In the 1990s, when artists such as Bryan Adams or Wet Wet Wet were sitting aloft at the top of the charts for 10 or more weeks, this made actual TV news. Nowadays, songs that remain as number 1 for several weeks don't trouble the mainstream news in the same manner. But chart history in the UK was changed irrevocably in March 2017 (the same time as the Monkman vs Seagull *University Challenge* clash between our Cambridge colleges, Wolfson and Emmanuel), when 16 songs from an album mathematically titled *Divide* by ginger-haired singer-songwriter, Ed Sheeran, occupied all but 4 of the Top 20 singles charts positions! Some felt that this made a mockery of the charts and it was all but impossible for new and upcoming artists to break through to mainstream attention.

The Official Charts Company now limits the number of recorded streams at 10 plays per person. Further, as of July 2017 they have placed an artificial limit of artists being only allowed a maximum of 3 songs in the Top 100 charts. For songs that have been in the charts for a lengthy period, the ratio of streams to count as 1 sale has been increased to 300:1 (compared to the 150:1 for newer songs). Hence, the longer a song is in the charts, the faster it will drop out of the Top 100. In theory, this should result in more artists having exposure in that coveted Top 40. After the death of Prince in 2016, 6 of his songs appeared in the Top 100 – something that would no longer be possible with the new rules. So, if any iconic music acts such as ABBA, Bob Dylan or Bruce Springsteen were to pass away, current rules would mean that they would not dominate the charts in death as they did in life. And going forward from here, it will be a continual battle between the mathematicians behind the Official Charts rules and the musicians who aspire for global dominance.

So, you might be thinking, 'OK, the stats men and women behind the charts might influence the rankings, but surely they don't actually influence the content of the music. The musicians will still continue to produce their music, and as an end consumer, it doesn't affect me.' You would be wrong unfortunately. The nature of the streaming industry and the battle for success there has changed the actual output and shape of pop songs.

Think back to some of the classic pop power ballads and epic songs of the 1980s such as 'Hotel California' by The Eagles or even 'Eye of the Tiger' by Survivor. These songs have long introductions of around a minute before any singing begins. Put on chart music now and you'll find most songs feature opening lyrics within seconds, such as Clean Bandit's 'Rockabye' vocal starting after just a second! The reality is a little bit more nuanced but the data backs this up.

Research conducted by Ohio State University doctoral student Hubert Léveillé Gauvin found that intro lengths to songs had dropped by 78% between 1986 and 2015, from more than 20 seconds to now only about 5 seconds. What does the nature of the streaming industry have to do with this? Well, if a song is played for less than 30 seconds on Spotify, it doesn't count as a play and hence doesn't contribute towards its weekly streaming numbers. So there is pressure on musicians to get consumers hooked onto songs fast, to prevent them skipping to something else – a musical 'survival of the fittest'. It's vastly easier to skip tracks now than it used to be in the past – a quick click compared to having to fast forward an analogue tape – and that has influenced how songwriters and producers think when penning future hits. People also listen to playlists now as much as albums, so one song may be your only chance to get a 'play' before the listener moves onto another artist. These formulae behind streaming charts and Spotify's '30-second rule' may bring a premature death to the epic song intro that was so widespread in the 1980s. The physical format, and the maths behind the rankings, really do change the content.

The process of writing music is considered a creative one. Some writers sit there for hours fine-tuning their melodies and lyrics, while others sometimes find melodies emerging to them like a dream. Can you truly create new music?

Back in 2013, the top-selling song in the UK (and possibly the world) was the 'Blurred Lines' collaboration between Robin Thicke and Pharrell Williams. However, in 2015, a court verdict ruled in favour of Marvin Gaye's family, who found that 'Blurred Lines' had copied Gaye's 1973 hit 'Got to Give It Up'. As well as receiving £5 million in damages, the family would also receive 50% of any future royalties. If you listen to both songs, you can definitely hear similarities. Williams claimed in court that Gaye's music was part of the soundtrack of his youth, and acknowledged a likeness between the songs though he was only 'channelling that late-70s feeling' when he co-wrote the song. In 2017, Ed Sheeran settled out of court for a £14 million copyright infringement claim against him in America over his song 'Photograph'. This time, songwriters Thomas Leonard and Martin Harrington claimed that Sheeran's ballad had a very similar structure to their song 'Amazing'.

Musicians draw inspiration from each other. But this brings us to a question about the uniqueness of music and whether we might one day run out of new music. Let's narrow that question down and ask whether all of the good melodies have been used up? Or more pertinently, have all the good, *pleasant-sounding* melodies been used up? What are the chances of 2 songs having the same melody, but being independently created?

Let's start with something simple, such as 'Three Blind Mice'. How many 3-note songs are there? If we use the regular musical scale, there are only 7 possible notes to choose from. If there were 10 notes (labelled from 0 to 9), we could work out how many different 3-digit permutations existed. Mathematically, a permutation is different from a combination as the order of digits matters in the former: 123 is a different permutation to 321, but they are both the same combination as they contain the same digits just in a different order.

For those that are fans of British comic double act Morecambe and Wise, one of the finest sketches in 1971 involved André Previn, the world-renowned conductor and composer who won 4 Oscars for his film work. Eric Morecambe begins playing the piano, purportedly performing Edvard Grieg's 1868 majestic piano concerto in A minor. However, Previn looks frustrated and tells Eric that he is playing 'all the wrong notes'. Eric ripostes, 'I'm playing all the right notes – but not necessarily in the right order.' So Eric could have claimed that his music was correct as a combination, as it contained the right notes, but not in the correct order.

So back to the mathematics of permutations. If we had 10 notes, labelled from 0 to 9, that could be used to represent a 3-note song, the smallest number would be 000 and the largest would be 999. This gives us a total of 1,000 permutations. Mathematically, we can quickly calculate this by $10 \times 10 \times 10 = 10^3 = 1,000$.

But we only have 7 notes to choose from, so instead of 10^3, we calculate $7^3 = 343$. Of course, I'm being simplistic here. There are not just 7 notes, in the same way there are not just 7 colours in the visible spectrum – there's an entire range in between. The visible spectrum is a continuous range of light with different wavelengths. The same analogy is useful for sound. Musicians can also break down sounds in different sorts of musical scales, and you can have 12 notes in a chromatic scale. With those 12 notes, we can deploy the same method to give us $12^3 = 1,728$ different permutations of 3 notes. Some of them are likely to sound just as memorable as 'Three Blind Mice'.

Mathematics can offer us a way of looking at music as separate notes, arranged in a particular order to please our ears. Music is a way of humans expressing themselves in an auditory manner (and physical if you include the performance). We can see music from the perspective of the consumer (listener), creator (composer) or even the player just performing the notes written down by a composer. Not everyone creates music, but the vast majority of us will listen and some of us will play based on other compositions. Mathematics is not a million miles from this. In mathematics, only some of us create new maths, but the vast majority are consumers of this and can still appreciate its importance.

Chapter 8 Puzzle

Eurovision Madness

For me, the Eurovision Song Contest is a musical highlight of each year, though I yearn for the day that the Jedward twins return back to the show for Ireland. Entries from France, Germany, Spain, Italy and the United Kingdom are guaranteed places in the final every year.

With the following song choices, why might only the UK entry get the infamous nul points?

France's song is 'Born a shining star'.
Germany's song is 'People in the crisis hour'.
Spain's song is 'Time out never ends'.
Italy's song is 'Rock it fallen friends'.
UK's song is 'New undying love'.

CHAPTER 9

'Fit for Maths, Fit for Sport'

The Maths of Fitness

'If it's nice, do it twice!' 'What's my name? Sergeant Pain!'

Whenever I'm back in my hood of East Ham, you can find me on Saturday between 11.45am and 1.15pm, in a sweaty heap, deep into a circuits class at East Ham Leisure Centre. And those are the words that come from our charismatic instructor, Dave McQueen, when the class needs picking up after our latest set of press ups, triceps dips or jumping jacks.

I love taking on multiple different projects at the same time, which can mean I am often squeezed for time. When I retire one day, I can imagine my life being consumed by reading and an abundance of sport. But for now, while I love my sport, during the week I tend to try to optimise what little time I have. On some days before teaching at school, I'll spend 4 minutes (yes 4!) doing exercise before hopping in the shower. And when I say exercise, I really do mean it. It's 4 minutes of HIIT – high intensity interval training, though without the inter-vals! The rapid raising of the heartbeat and maintaining of that rate keeps me ticking over during the week till my Saturday circuits-whipping time.

Also, when I'm in Cambridge, I try to run with the university middle-distance running club, the Cambridge University Hare & Hounds (despite its suggestive name, it has no connection whatsoever with fox hunting!). It's an incredibly friendly club, ranging from very serious runners, to regular local runners such as me, and even to those who are jogging for the first time. It was at my running club that I first heard the term that best describes my sporting ability: NARP. A 'non-athletic regular person' is someone who likes to play sport and keep in shape but possesses no extraor-dinary sporting talents and is by no means elite. And that is me, a NARP – though one with a keen sense of numbers, and using the efficiency of intense workouts to maximise what I have. If I characterise my sporting abilities as a dog, I would be a feisty terrier: not a big dog, but one that has high energy levels!

One thing that chimes with me about running or circuits classes (or any sport to be honest) is that numbers can play a crucial role. Of course, the endorphins rushing through your body are a kick but I also love using numbers as targets to stretch myself – whether it's getting to that cleanly executed tenth press up or completing another 400-metre track repetition in 80 seconds (yes this sounds slow, but is challenging for a mortal NARP such as me to do a few times back to back!).

The fitness industry is serious business. In the UK alone, the total market was worth £4.7 billion in 2017. You flick through the magazine section of most stationery shops or supermarkets and you'll find the front cover of health or running magazines with headlines such as 'Run a 5km in under 25 minute in 5 weeks', 'Gain 10kg of muscle mass in 10 weeks' or 'Reduce your weight by 10% in a month'. As the fitness industry continues to grow in prominence, so will experts who use sports science and numerical targets to show you how their workouts and routines will get you in the best shape ever.

In the more innocent era of the early 1990s, Derrick Evans (known to the British population as Mr Motivator) would don his fluorescent spandex in front of millions on breakfast TV, encouraging us to follow his routines. By the late-2000s we had Shaun T from the US shouting at us through his *Insanity* DVDs to 'come on, y'all' and keep up with his high intensity exercises. More recently in the UK, Joe Wicks – whose Body Coach YouTube videos have seen him become an almost household name – has been whipping young professionals into shape. Joe is an idol in my family!

The fitness industry is about coaxing us into our best possible shape. And to do this requires us to make sacrifices and undergo intense training. Building a good sense of numbers can help us reach our goals more quickly.

A few years back, I used to enjoy doing the 5 kilometre local Parkruns when I could get up in time for a 9am start on a Saturday. I haven't been much as of late; when I have, I don't have registered times as I've stopped bringing the necessary barcode. For those who don't know, Parkrun is a global phenomenon, with around a thousand free, timed runs held on Saturday (or sometimes Sunday) mornings. My local club in east London is the purple-clad East End Road Runners. At my peak, I moved around my local Beckton 5,000-metre course in 18:43 and completed a 10-kilometre race in 39:56 (I remember sprinting those last few strides in order to get under that psychological 40-minute barrier) – to put in context how quick elite runners are, Mo Farah won his London 2012 gold medals for the 5,000 metres in 13:41 and 10,000 metres in 27:30 (and yet still, these are more than a minute behind the mind-blowing world record times of 12:37 and 26:17).

My family, who think of me as a runner, often ask me when I will be running a marathon. The answer is I'm not sure. I've completed a couple of half marathons, in 2013 getting round the Bedford course in just over 1 hour and 28 minutes – though my lack of training meant that this was pushing myself too far and my East End Road Runners club's minibus on the journey home had to endure the contents of an unpleasant regurgitation! My family have said that if I've done a half in 1:28:00, then surely I can get a full marathon done in under 3 hours. Unfortunately, linear extrapolation of data does not quite work with sports.

The marathon is seen as one of the benchmarks for an amateur runner. Once you join a local running club, you've inadvertently joined a treadmill (pun intended) to eventually complete a marathon. Even for elite distance runners who dominate 10,000 metres on the track, a marathon is seen as the ultimate distance race. (Of course, there are those ultra-marathon runners who take it to the next level.) Mo Farah retired as a multiple Olympic and World Championship gold medal winner on the track to focus the final part of his career as a marathon runner.

On 6 May 2017, there was an attempt by Nike to achieve the holy grail of distance running: to complete a marathon in under 2 hours. The 'Breaking2' project aimed to find runners who could complete 26 miles and 385 yards, or 42.195 kilometres under that psychological time barrier. In front of primary school teachers and some primary students in July 2017, I actually did a speech about how to dream the impossible, and this marathon effort was my vehicle for this talk. As well as the athletes drawing on the reserves of inner strength, this Breaking2 effort would require science and sport to coalesce to give the athletes the best possible opportunity. Nothing would be left to chance.

Just to give a sense of the scale of the attempt, here are some numbers. The current world record for the men's category is 2:02:57 by Kenyan Dennis Kimetto recorded at the 2014 Berlin marathon. If you do the maths, we get a pace of 4:42 minutes per mile, back to back for the entire 26.2 miles. Convert this to pace in metric measurements, we have 2:55 per kilometre, which equates to 70 seconds per lap of an athletics track or 400 metres. If you can remember back to your school days, 70 seconds round a track is a decent time. Try doing this 105 consecutive times! Even when you break it down to 100 metres, it would require running a 17-second race. Then try repeating this 421 more times! Hopefully the sense of the challenge awaiting these runners is dawning on you.

Back to the race attempt. Some of the best runners were chosen – including the current Olympic champion, Kenyan Eliud Kipchoge, as well as the half marathon world record holder, Eritrean Zersenay Tadese. Nike selected 30 of the best runners on the planet to serve as pacers for the race. The course itself was selected for its low altitude, calm conditions and short lap lengths – Formula 1's Italian Monza track. The pacemakers even positioned themselves to shield the athletes from wind resistance. Nothing was left to chance. Biomechanics, nutrition and shoe technology were pushed to their boundaries for this moon-shot attempt.

The runners started at 5:45am with Ethiopian Lelisa Desisa dropping off from the pack after 30 minutes (finishing with 2:14:10) and Tadese falling behind the pace at mile 20 (finishing with 2:06:51). Only Kipchoge remained but he too could not make it, just missing his target by 25 seconds with 2:00:25, just under 1 second too slow per mile! While this smashed the official world record of 2:02:57, it did not count as an official record as they infringed several official rules, including the fact you are not allowed pacers. However, what we learned is that science and maths combined with human endeavour can push athletes to their limit.

Gyms in late December can seem like a morgue, empty and disused. But by early January, they are like the sardine cans of jammed London Underground tube trains. Gyms are frequently crowded and congested in the new year, when people have set new year's resolutions. One of the most common targets is a weight loss programme, where an adult may say they want to lose 10 kilograms in a year or a certain percentage of their current level. Weight loss can be a useful measure, but by itself it can be meaningless. Weight loss caused by a drastic reduction in calorific intake does not necessarily boost our health and fitness. So, are there are other measures of fitness?

An alternative measurement is the Body Mass Index or BMI. This was devised in the 1840s by Belgian polymath Adolphe Quetelet and has been used for more than a century by health professionals to assess whether a patient is over- or under-weight. It aims to estimate if an adult has a healthy weight by dividing their weight in kilograms by their height in metres squared.

Most health experts would use the following criteria for BMI:

<18.5 underweight
18.5–24.9 normal weight
25–29.9 overweight
>30 obese

Higher BMI is associated with higher risk for certain diseases such as cardiovascular disease, high blood pressure, type 2 diabetes and even certain types of cancer.

When BMI was first devised, Quetelet wanted a simple 'rule of thumb' measure to assess health. In modern times, we have technology to deal with more complex calculations. BMI does not take into account the 3-dimensional nature of our physicality, nor does it differentiate between muscle and fat. We could take the example of a couch-bound sports fan, who is 1.83 metres (6 feet) and weighs 92 kilograms. This person would have a BMI of 27. By contrast, an elite sportsman may be the same height but a few kilograms heavier and hence have a BMI of greater than 27. Would our assessment be that the elite sportsman is more overweight than our couch fan? I think not, especially as we know that muscle weighs more than fat.

In 2013, Nick Trefethen, a professor of Numerical Analysis from the University of Oxford, wrote in to the *Economist* magazine critiquing BMI:

The body-mass index that you (and the National Health Service) count on to assess obesity is a bizarre measure. We live in a 3-dimensional world, yet the BMI is defined as weight divided by height squared. It was invented in the 1840s, before calculators, when a formula had to be very simple to be usable. As a consequence of this ill-founded definition, millions of short people think they are thinner than they are, and millions of tall people think they are fatter.

It is still slightly alarming that BMI is used as a measure by many in the health industry. Looking at the formula, we have height squared, or an exponent/index of 2. Our world is 3-dimensional and we know that individuals do not scale in a linear fashion. Perhaps a compromise index of 2.5 might work better. Interestingly, Quetelet did write in 1842 that 'if a man increased equally in all dimensions, his weight at different ages would be the cube of his height'. So there is an acknowledgement about the use of the square being possibly flawed.

Obesity is going to be a burgeoning issue in the twenty-first century and if societies are placing value on BMI, then perhaps a more nuanced discussion needs to occur on the use of this statistic. Other measures to see if an adult is overweight are waist circumference or the waist-to-height ratio. Scientifically, a person with excess weight around their abdomen has a higher risk of heart disease and other metabolic diseases (possibly as fat impacts internal organs such as the heart and liver).

Maths can be a pragmatic tool to calculate our short- and long-term goals, and can also measure our growth in fitness levels. In the UK, the National Health Service encourages adults to take part in the 'Couch to 5k', a running plan for beginners to move from sedentary to active lifestyles. It is based on weekly quantitative plans that can help motivate runners to move towards their goals.

In science school lessons, I recall using heart rates as a measurement. First, we measured our resting heart rate, then after performing some exercise (usually a lap around our concrete school playground), we measured it again. One measure of fitness is how rapidly your heart rate returns from an elevated level to its resting one.

Sport is a field where marginal gains can make a significant impact on performance. Prior to 2018, the England national men's team had never won a World Cup penalty shoot, indeed having the worst record of any country. We had lost 4–3 to West Germany in 1990, 4–3 to Argentina in 1998 and 3–1 to Portugal in 2006. England manager Gareth Southgate himself missed a penalty in the 6–5 shootout defeat to Germany in the 1996 European Championship. Realising the impact of marginal gains, Southgate ensured players looked at individual process and techniques. For example, in training, players walked slowly from the halfway line to imitate match conditions compared to their previous hurried kicks. They also practised 1 or 2 stock penalties executable under pressure and had a definite list and order of penalty takers. This attention to detail (combined with the player's steely nerves) resulted in England's first ever World Cup penalty triumph, a deliriously giddy 4–3 victory against Colombia.

Many coaches are inspired by *kaizen*, the Japanese term for 'improvement'. In Japanese business, those that follow *kaizen* try to continuously improve all functions. They never aim for perfection, focusing on the small rather than the big. This philosophy's concentration on progression results in compounding the improvements. As individuals what we can take away from this is that small improvements in our running technique or breathing, or even the type of trainers we wear, can result in significant improvements. Marginal gains are almost like the sporting version of the supermarket Tesco's 'every little helps' campaign, where the shops looked after the pennies, and the shoppers' pounds resultantly looked after themselves.

One thing that sport can demonstrate to us is that dedication and hard work can help counter any lack of natural predispositions – to a certain level at least. During the summer of 1995, I was mesmerised by the World Athletics Championships in Gothenburg, Sweden. Jonathan Edwards, a deceptively powerful athlete from Gateshead, hopped skipped and jumped his way into the record books. Jump number 1: 18.16 metres, just over 59 and a half feet. Jump 2: 18.29 metres, 60 feet (useful real-time conversion skills being demonstrated by the TV commentators between centimetres and feet). Both were consecutive world records, and both jumps sublime. Edwards was far from the tallest athlete on the course but it was his technique that propelled him to gold, and it's still the longest legal recorded triple jump in history.

How did these events inspire me? Well, I thought, I want some of that triple jump glory myself. So over the course of the year, before the annual school sports day, I prepared myself methodically for the triple jump.

I tried to understand what gave Edwards his world records. Technique, speed and, I assume, a lot of practice. I watched Edwards's jumps on videotape and tried to replicate them in my back garden with his use of his arms and legs in a particular manner. Triple jump only required short bursts of speed – so I practised this as well over the course of the year, focusing on 10–15 metre sprints.

Come Sports Day, I was ready. I'd like to say that the shortest boy (me) in his Year 7 field came first, but it wasn't quite that. I did come second, beating boys who were significantly taller than me, but it demonstrated how technique and repetition can overcome others with more natural gifts. When I repeated the triple jump in Year 8, I dropped down the field to third place, as my technique could no longer compensate for the limits of my physicality as other boys started to grow even more beanstalk like compared to me.

Mathematics can be like my triple jump story. There may be others who perhaps seem to have a natural affinity. But I genuinely believe that with deliberate and targeted effort, all of us can achieve more than we expect of ourselves.

'The limit of human endurance. If you find it, call us.' There has been a Nike poster up on my bedroom wall for perhaps a quarter of a century, zooming in on the face of a non-descript athlete, perhaps exhausted and at a finish line. The use of maths can help propel us towards personal fitness goals but also show us that, sometimes, numbers are just targets for us to smash.

'FIT FOR MATHS, FIT FOR SPORT'

In a world focused on data, the use of personal smart-phones and measurements, we are in a paradoxical era where more people than ever are pushing the limits of their amateur fitness but we also witness record levels of western obesity.

Chapter 9 Puzzle

Joe Wicks's Mathematical Workout Routine

I'm a fan of YouTube fitness sensation the Body Coach (aka Joe Wicks). Joe Wicks often sets his YouTube routine for his millions of followers while on his travels around the world. On his trip to New York to set up an outdoor workout in Central Park for his American fans, he visits the National Museum of Mathematics, known as MoMath. In homage to visiting a mathematical museum, Joe sets up the following body-weight only routine to be performed continuously for 17 minutes on repeat. (Of course, Joe would pick a prime number for the routine as he wants his followers to be in their physical prime!)

$2 \times$ Burpee
$3 \times$ Calf Raise
$12 \times$ Lunge
$13 \times$ Mountain Climber
$16 \times$ Press Up
$19 \times$ Squat

How many tuck jumps should I perform to fit in with this routine?

CHAPTER 10

'Money, Money, Money'

The Maths Behind Everyday Finance

I find beauty in mathematics. To adapt an early nineteenth-century French art slogan, 'maths for maths' sake', should be sufficient. This expresses a philosophy about the intrinsic value of understanding the patterns in the world around us, divorced from any utilitarian function. However, all maths teachers have to be ready for the moment when a child is befuddled in a lesson and raises their hand, not for assistance but for a juncture of existential angst:

'What's the point of learning maths?' This is a question that is hurled at me in various guises during maths lessons. And of course, while I find maths fascinating for its own sake, young people in particular often want to see a direct relevance to their future lives, to understand why they should spend time learning it.

Whether you are rich or poor, you will need to consider money at some point in your life. And as numbers are the language in which is money is expressed, financial mathematics comes into play. From relatively straight-forward calculations of simple interest to the 'weapons of math destruction' that contributed to the financial crisis in 2008, maths is simply too important for us to ignore.

I am an ambassador/supporter for the charity National Numeracy. Founded in 2012, their raison d'être is to challenge poor numeracy in the UK, particularly acute within adults. Just shy of 50% of adults in the UK have the numeracy skills we would expect from an 11-year-old in primary school! Admittedly, what 11-year-olds learn in a school maths curriculum is different to the daily numeracy demands on adult life, but this is nonetheless a shocking statistic. As a taster of how much we struggle with numbers in this country, the following question ably demonstrates our mathematical woes.

If you are on £9 an hour, what would be your hourly rate become if you were offered an increase of 5%? With or without a calculator, half of this country would show significant bother in working out that the new hourly rate would be £9.45 (10% of £9 is 90p and 5% is half of that at 45p, which is then added back onto the £9 rate).

Part of the obstacle facing the UK in mathematics is our cultural and attitudinal issues about the subject. Often when I introduce myself to a new group of people, in a pub or at a dinner party, and say that I'm a maths teacher, a hand will be raised. Someone will announce that they couldn't do maths at school. Then other hands will follow, all agreeing that they were rubbish at maths. Being bad at maths is seen as a badge of honour.

I will of course concede that there is a small minority of people (3 to 6% of the population) with varying degrees of dyscalculia, a condition that negatively affects the ability to acquire mathematical skills.

The charity National Numeracy came to the forefront of the news when they forced beauty giant L'Oréal Paris to alter an advert that had included a throwaway boast about not being good at maths. In a 2015 ad, British actress Dame Helen Mirren said, 'Age is just a number. And maths was never my thing.' Mike Ellicock, the charity's CEO, tweeted their distaste that 'Throwaway remarks about being "no good at maths" are so easy to make and so damaging in the way they normalise negative attitudes'. To their credit, L'Oréal Paris acknowledged their error and removed their comments.

This highlights the challenge that maths has, to overcome historical and long-lasting public negativity. On 16 May 2018, the UK celebrated its first ever National Numeracy Day. We had such positive coverage across the board: articles in mainstream press such as the *Financial Times* and London's *Evening Standard*; radio interviews including BBC Radio 4's flagship *Today* programme and at one stage we were on *BBC Breakfast News* (I was alongside *Countdown*'s numbers whizz Rachel Riley) and ITV's *Good Morning Britain* (with Money Saving Expert's Martin Lewis). However, this is just the start.

Most people who had scarring school experiences will often remember back to struggles with basic mathematics such as the 4 operations of maths (addition, subtraction, multiplication and division). Let's be straight: this is much more about numeracy, the day-to-day number skills we need to navigate the world.

Mathematics is much more than numeracy, though numeracy is a part of maths. To handle the demands of being a modern citizen, we need to be financially responsible, and understanding the maths behind finance can be a starting point.

The most fundamental financial store of our savings is a bank account. Most of my school students who are aged 13 and above should have seen the terms *simple* and *compound interest*. As a thank you for keeping money deposited in a bank, you will receive interest payments over a period of time. Likewise, if you borrow money from a bank (or any other institution), you will pay back interest.

Interest is essentially the cost of borrowing the money – what you will pay to the lender for providing the loan. Normally this is expressed as a percentage of the principal (the original amount of the loan or deposit). We hear of saving accounts with a 3% annual rate, or 5% for a mortgage.

Mathematically, there are 2 types of interest: simple and compound. Simple, unsurprisingly, is straightforward. From my experience, though, simple interest is not particularly common in the real world apart from in the heads of GCSE exam question setters (so it's still something most 16-year-olds in the UK need to master). I have seen some car loans with simple interest, where interest is calculated on the principal loan balance on a daily basis. Here, payments are first applied to any interest due, and then after that to the principal balance.

Let's look at an example of simple interest. Say I take out a loan for a silver blue Nissan Micra (which may or may not be the car that I share with my mother) for £3,000. The daily interest amount is equal to the annual rate (let's say 5%) divided by the number of days in a year (365 for non-leap years). On a £3,000 loan at 5%, the interest daily would be £0.41 (£3,000 × 0.05 ÷ 365). This means that essentially I pay a fee of 41p every day to have this loan. And I have to pay that 41p back each day before I can pay off the £3,000 I have borrowed.

Simple interest loans are usually paid back in equal monthly instalments. The financial term for this is to say that a loan is 'amortising', which means that a part of each payment contributes to pay down the interest. The balance of it is applied to the principal loan balance.

My dream car is a Mini – smart, compact and energetic – which fits me and my personality perfectly! I fell in love with the Mini when I was mesmerised by the iconic car chases of Michael Caine and his gang's shenanigans in the 1969 thriller *The Italian Job*. However, the original British Minis are manual cars, and sadly I'm not a fan of driving with manual gears or 'using a stick' as my American friends would say. So currently, my heart is set on buying an automatic Mini Cooper 3-Door Hatch in the next few years. It's on the market first-hand for about £20,000. In reality, I would probably purchase a good quality second-hand one, though for this exercise let's look at the numbers behind getting the new one with a car finance deal.

On the Mini website, you have the option of making monthly payments over 4 years (so 48 months). I'll put down an upfront deposit of £3,000 leaving the principal outstanding of £17,000 with a theoretical annual interest rate of 4%. My monthly repayment is £383.84. How is this calculated? (Note that I've initially completed these calculations on an Excel spreadsheet, so I haven't rounded any figures right till the very end.)

Step 1: Convert my loan's annual interest rate to a decimal number. So 4% is 0.04.

Step 2: Divide this number by 12 months to find a monthly interest rate. So $0.04/12 = 0.0033$.

Step 3: Multiply this amount by the total amount of my loan: $0.0033 \ldots \times £17,000 = £56.66$.

Step 4: Add 1 to my answer from step 2.

$1 + 0.0033 \ldots = 1.0033 \ldots$. 1 represents the principal amount borrowed and the 0.0033 is effectively the interest rate.

Step 5: I take the answer from step 4 and raise it to the power of how many months I'll make monthly payments to show how this payment compounds over 48 months: $1.0033^{48} = 1.173199$.

Step 6: Take the reciprocal of step 5. That is $1/1.173199 = 0.852 \ldots$ This shows the original amount borrowed as a proportion of total repayment (which includes the interest).

Step 7: Subtract the answer of step 6 from 1. So 1−0.852 ... =0.147 ... This shows the percentage of interest throughout the life of the borrowing.
Step 8: Step 3 divided by step 7 = £56.66 ... /0.147 ... = £383.84.

So what the numbers tell me is that a 4% loan for a 4-year period costs £383.84 monthly. At the end of the 4 years, I would have paid £18,424.52 in monthly payments. If we include the initial £3,000 deposit I made, the real cost of my Mini dream, including all of the interest on the loan, would be £21,424.52. So it would cost me an extra £1,424.52 in interest to buy the car with a £17,000 loan, rather than buying it outright with a payment of £20,000.

In the financial world, there is a technique called sensitivity analysis which investigates what happens to the output (our monthly repayments and total cost) when we alter the inputs (interest rates, deposit amount, repayment period). So instead of a 4-year loan, if I stretched that out to a 6-year loan, my monthly repayments would drop to £265.97. Sounds good? A more affordable monthly payment, sure, but at the cost of a higher total cost. Including the initial £3,000 deposit, the total cost for my Mini would be £22,149.84.

The interest rate offered on the loan has a significant impact on the number crunching. Instead of a 4% car loan, let's say that I'm offered 6%. Doesn't sound like a noteworthy increase. But on the original 4-year deal, this increases the monthly payment to £399.25 (from

£383.84), which ramps up the total monthly repayments to £19,163.73. So this Mini would cost an additional £739.68. Although it didn't sound like a big increase, that's actually a lot more money to spend.

What can we learn from this situation mathematically? When choosing to enter a car loan, there is always a trade-off. If you're operating on a tight financial budget, then a lower monthly bill is appealing – though this means that you'll have more monthly payments to make and ultimately pay a higher overall price for your dream (or not-so dream) car. If you want to forge a way out of debt as quickly as possible then, of course, you'll want to pay as high a deposit as possible and increase your monthly repayments.

'Compound interest is the eighth wonder of the world. He who understands it, earns it ... he who doesn't ... pays it.' 'Compound interest is the most powerful force in the universe.' 'Compound interest is the greatest mathematical discovery of all time.' These mighty powerful quotes are often attributed in various forms to the colossus of physics, Albert Einstein.

The most typical compound interest scenarios we see in the real world are with deposits in bank accounts. Here, the interest is added to the deposit (principal) after a particular period, usually a year. So for next year's calculation the interest is worked out on a larger amount of money than was previously invested.

We can compare 2 banks, the Simple Interest Bank against the Compound Interest Bank, both offering 5%

per year. Let's say I've had a very successful accumulator football gamble that's come through. The final result is West Ham securing an unlikely victory away to Manchester City, netting me £500 in profit. I invest this £500 in both these banks for 3 years.

At the Simple Interest Bank:
Principal at the start = £500
Interest for 1 year = 5% of £500 = (5/100) × £500 = £25
Interest for 3 years = £25 × 3 = £75
Total in bank after 3 years = £500 + £75 = £575

This can be expressed in this formula P × R × T where:
P (principal) is the amount deposited
R is the rate of interest per year
T is the time in years

At the Compound Interest Bank:
Principal at the start = £500
Interest in the first year = (5/100) × £500 = £25
Principal after 1 year = £500 + £25 = £525
Interest in the second year = (5/100) × £525 = £26.25
Principal after 2 years = £525 + £26.25 = £551.25
Interest in the third year = (5/100) × £551.25
= £27.56
Principal after 3 years = £551.25 + £27.56
= £578.81

So the Compound Interest Bank gave us £578.81 compared to £575 in the Simple Interest Bank after 3 years. More, but not by much. The grandiose power of compound interest starts to take shape after we start increasing the years. We can use mathematical indices to make our calculations easier. For 2 years: $1.05^2 \times £500 = £551.25$; for 3 years: $1.05^3 \times £500 = £578.81$.

These are principal amounts I would have with the Simple Interest Bank vs the Compound Interest Bank after the following periods.

After 10 years: £750 vs £814.45
After 20 years: £1,000 vs £1,326.65
After 50 years: £1,750 vs £5,733.70
After 100 years: £3,000 vs £65,750.63

A small difference starts becoming gigantic. Comparing savings accounts with an annual rate of 3% or 5% sounds trivial, but the difference over a 10-year period of compound interest is startling. It is the difference between a £10,000 investment leaving you with a gain of £3,393.16 or £6,288.95. Compound interest is like a snowball falling down a ski-slope, picking up snow along the way, and becoming a deadly ball able to cause avalanches. And this is something we should always heed in life: the power of things to burgeon through the power of compound interest. Of course, this is most evident in the financial world. Far too many people misunderstand this concept when borrowing money from payday loan companies or equivalents and end up owing them astronomical sums when they've only borrowed a small amount. We need a country that is more financially astute and compound interest under-standing is essential for that.

Even in the world of self-improvement, compound interest may help you realise the challenges of some goal setting. Let's say on 1 January, you aim to become better by 1% each day over the course of the year. This doesn't involve simple interest, but a compound growth concept as you have a new 'principal' amount at the end of each day. After 1 day, you are 1% better. After 2 days, you are 2.01% better than at the start of the year. After 10 days, you are 10.5% better. After a month of 30 days, you are 34.8% better. And after a year of 365 days, if you somehow manage to keep on improving 1% daily, you are now 3,778% better or 37.78 (1.01^{365}) times better than you were at the beginning of the year!

One of my favourite songs to jig along to is 'ABC' by The Jackson 5 band that was released in 1970 and led by the all-singing-and-dancing child virtuoso Michael Jackson. The song went 'A B C, it's as easy as 1 2 3'. If only they had ventured a couple of letters further, to D and then to E, we might have had an opportunity to expose dance-floors of weddings and birthday parties to the mathematical concept of 'e'.

All school students will have been exposed to π (from the area of a circle). However, e is something that students only encounter during post-16-year-old maths. The e stands for exponential and was first discussed by the Swiss mathematician, Jacob Bernoulli, in 1683. (The Bernoulli family, like the Jacksons, were astonishingly accomplished, albeit in maths.) The exponential occurs in problems about compound interest, led to the development of logarithms, and is handy for mathematicians to understand how variables like temperature, radioactivity or even how populations of species (like us humans) increase or decrease.

Leonard Euler, another Swiss mathematician, was the eighteenth century's mythical rock star of maths and found a way of linking e, i (an imaginary unit number, a square root of a negative where $i^2 = -1$) and π:

$$e^{i\pi} + 1 = 0$$

This is known as Euler's identity and is either considered an example of the most beautiful equation in maths, expressing a profound connection between the most fundamental numbers in maths, or like England's form at most men's international football tournaments, this is the most overrated and overhyped equation, as some of my very clever maths friends at Cambridge muse.

Back to the letter e, and this time let's say you've got £100 to deposit. You deposit that £100 in a bank and they will pay 5% interest per year in a compound manner. We now know how to calculate what this would be after a period of 5 years. You would have £100 × $(1 + 0.05)^5$ = £127.63.

Let's say a bank is generous enough to give you an interest rate of 100% annually (imagine a Ponzi scheme fraudster or an Icelandic bank pre-2007). If you deposited £100, you would have £200 at the end of the year.

Let's then imagine a second scenario where the bank gives you *half* the interest rate, but for *double* the number of periods, so a 50% interest rate twice in a year. After 6 months, you would have £150. Then at the end of the 12 months, you would have £150 × 1.5 = £225. This is clearly more than 100% paid in one lump sum. The calculation is as follows:

After 1/2 period: $(1+1/2)^1 = 1.5$
After 2/2 period: $(1+1/2)^2 = 2.25$

As you are investing £100, the amounts after each period would be £150.00 and £225.00.

Further still, let's picture a third scenario where the bank splits the 100% into 3 equal parts (33.33%) and gives interest every third of a year (every 4 months) instead. Your calculation would be as follows:

After 1/3 period: $(1+1/3)^1 = 1.3333$
After 2/3 period: $(1+1/3)^2 = 1.7777$
After 3/3 period: $(1+1/3)^3 = 2.3703$

As you started out with an investment of £100, the final sum in your account would be £237.03.

We can see that increases in the frequency of compounding results in you having more in your bank at the end of the year. You might now ask: if we could increase the period of compounding to 3 months, 2 months, 1 month or even to smaller gaps such a half day, an hour, a minute, a millisecond and on, would we eventually get to an infinite amount?

If we apply the interest rate continuously, the pattern if we divide the period into n equal parts with an interest rate $1/n$, will be the following at the end of the period: $(1+1/n)^n$

Let's look at a table where we look at n as we make n extremely large (rounded to 10 decimal places):

n	$(1+1/n)^n$
2	2.2500000000
3	2.3703703704
4	2.4414062500
10	2.5937424601
100	2.7048138294
1,000	2.7169239322
10,000	2.7181459268
100,000	2.7182682372
1,000,000	2.7182816925
10,000,000	2.7182816925

As n tends to become very large and closer to infinity, the limit of $(1+1/n)^n$ gets closer and closer to a fixed number, which is approximately 2.718. This is what mathematicians call e.

If someone declines to appreciate the elegance of the exponential function e, I feel a bit of dejection for them, but the cold realities of managing personal finances are of material significance. If a consumer ever has to take out a commercial loan for any purchase, or make a decision about how to invest their savings, an understanding of the connotations of compound interest will make their lives considerably easier. When compound interest is understood, it can be your friend. When it is misunderstood, it is an enemy you do not want to cross.

Chapter 10 Puzzle

James Corden and Russell Brand on a Big Day Out at West Ham

James Corden and Russell Brand decide to visit West Ham's Megastore beside the London Stadium before a big FA Cup home match. The club have a '20% off all purchases' offer on this day. So James bags himself a sweet deal where he buys a new claret and blue home shirt and scarf for £64 and gives the scarf to Russell.

As well as appreciating that £64 is a square number of 8, James and Russell try to work out what the original price of the shirt and scarf were before the discount. The shirt costs £40 more than the scarf before the 20% was applied.

What were the original prices of the shirt and the scarf?

CHAPTER 11

'Greed Is Good'

The Maths of the Financial Markets

'How Bobby Seagull predicted the 2008 financial crisis'. Well, so said the headline of a *GQ* magazine article after my cheeky-chappy appearances on *University Challenge*. *GQ*, formerly known as *Gentlemen's Quarterly*, is an international monthly men's magazine that typically features interviews with people such as rap star Stormzy, actor Gerard Butler, sports presenter Gary Lineker and former Formula 1 champ Jenson Button. But in April 2017, they decided a maths teacher doing a master's degree was worthy of their attention!

What was *GQ*'s headline referring to? In the run-up to that infamous day of Lehman Brothers' collapse, Monday 15 September 2008, there were signs that things weren't quite right. Conventional financial market analysts would look at more obvious indications such as Lehman's share price or the Credit Default Swap price (this is sort of like insurance against a company going bust, so you could essentially take a punt on whether you thought a company's future existence was looking ropey).

For me, one of the minor perks of joining the hordes of a global financial behemoth was that I didn't have to buy my own stationery. There was a plenitude of paper, pens and multi-coloured highlighters to keep me going through till England's next World Cup trophy! However, in those lazy summer months of June and July 2008, the stationery cupboards stopped being replenished. I was even compelled to visit WHSmith to buy some of my own pens for work! So I took the most rational action any young 20-something trader could do in trying to deal with this impending doom ... I headed for the chocolate vending machines.

At Lehman, most staff would divert a small portion of their salary to their lunch vending cards, a bit like a direct debit (tax efficient, we were told). I realised that if the firm went under, then the money on my lunch pass would be pretty low down the pecking order in terms of creditor repayment scheduling. So shortly before our demise, I bought a jumbo-sized shopping trolley into work and depleted the vending machines of at least a couple of hundred pounds' worth of chocolates. Mars, Snickers, Lion bars, Maltesers, Galaxy, Crunchie. You name a chocolate brand and I had it. At first, my immediate colleagues did think I was being very melancholy in my outlook for Lehman, but come the day of the crunch, my shopping trolley collection would prove my chocolate nest egg!

So I hadn't quite predicted the financial crisis, otherwise I would probably have left Lehman before its implosion, but I had seen some leading indicators that things weren't going well. Many people around at the time of the great financial recession of 2008 will have their own memory or tale to tell. Even people who were far removed from the chaos of the financial trading floors were impacted by this crisis. To try to prevent it happening again, all of us would benefit from understanding or, at least, appreciating the significance of the underlying mathematics that led to the debacle. And, as with simple and compound interest, the maths behind finance is something that we deal with on a day-to-day basis.

But why should your average man or woman on the street care about what happens in the financial world? Aren't these simply the wheelings-and-dealings of high powered brokers, done to enrich themselves at the expense of the masses?

Well, most jobs ultimately depend on the health of the economy. The stability of the financial markets plays a role of paramount importance in the economy, in particular the availability of credit (set by a central bank's lending rate). We might not like it, but bankers making decisions about the allocation of capital trickles down to make an impact on our daily lives. So we need to make sure that we at least understand the economic news so that we can make sense of the numbers jumping out at us from our TV screens.

Would wider understanding of the maths of the economy have helped avert the financial crisis of 2008?

CHAPTER 11

Maybe, maybe not, but as responsible citizens, we must surely understand how our economy works.

> *The point is, ladies and gentlemen, that greed, for lack of a better word, is good. Greed is right. Greed works. Greed clarifies, cuts through, and captures the essence of the evolutionary spirit. Greed, in all of its forms, greed for life, for money, for love, knowledge, has marked the upward surge in mankind and greed, you mark my words, will not only save Teldar Paper, but that other malfunctioning corporation called the USA.*

These are the words of Gordon Gekko, the fictional banker depicted by Michael Douglas in the 1987 film *Wall Street*, for which he won an Oscar, and an allusion to the title of this chapter. This movie depicted the boom years of the financial markets in the 1980s. There are scenes set on the hectic banking floors, with multiple computer screens, brokers with 2 phones taking client orders for buying or selling. I'll be honest, the sense of adrenaline rush from that scene did inspire me to consider banking as my first graduate career.

In the mid-2000s, there was a sense of invincibility in the global financial markets. The world economy was ticking upwards in a seemingly annual procession to the heavens, credit was easy, and consumers in the western world were buying hi-tech and luxury goods at ever-affordable prices. As a young person doing a maths and economics degree, it felt that the fast-paced world of banking offered the best place to push my capabilities to the limit.

Another type of banker is the central banker, and in the UK the Bank of England performs this role. Founded in 1694 in the reign of King William III, the primary modern role of 'the Old Lady of Threadneedle Street' (named after its London home) is to oversee monetary policy. The Bank's Monetary Policy Committee meets 8 times a year to set the interest rate policy. Whether we are savers or paying mortgages, this rate has far-reaching influence. Unfortunately, many students leave school at 16 or 18 with little understanding of how economics impacts wider life. Thankfully, the Bank of England has started taking steps to redress this. I was at the Bank of England in April 2018, where the Governor Mark Carney and Chief Economist Andy Haldane helped launch a new free school resource astutely called 'econoME'. This resource aims to de-mystify economics for school students, and hopes to complement the school curriculum.

The most basic of economic mathematical equations was referred to by Mr Micawber in Charles Dickens's novel *David Copperfield*:

Annual income twenty pounds, annual expenditure nineteen [pounds] nineteen [shillings] and six [pence], result happiness. Annual income twenty pounds, annual expenditure twenty pounds ought and six, result misery.

This is a simple equation for happiness which says that we shouldn't spend beyond our means, as individuals, organisations or even nations. Of course, there are times that we will borrow to smooth over tough periods but this is with the expectation that we can afford to pay back.

When a government injects spending into a project, let's say £10 million into affordable housing, the final increase in our nation's Gross Domestic Product (GDP) is likely to be more than £10 million due to the multiplier effect. As a reminder, GDP is a measure of the size and health of a country's economy over a period of time and is usually calculated as the sum value of the goods and services produced in a country. Regarding multipliers, there are 2 main types in economics: the money multiplier and the fiscal multiplier. (And before professional economists send me letters by the score saying I'm oversimplifying, I'll concede I'm going to provide a basic version just to provide a sense of what happens.)

Money multipliers are to do with fractional reserve banking. This is where banks will accept deposits, make loans or investments, but they are only legally required to hold reserves equal to a fraction of its deposit liabilities. This is why if every customer wanted to withdraw their cash from a major retail bank, it would not be possible (and hence the almost apocalyptic images of people queuing in their masses to withdraw their physical cash from British bank Northern Rock in 2008).

The key formula is the money multiplier $m = 1/R$ where R is the reserve requirement. If there is a reserve ratio of 10% (so a bank has to hold 10% of all its deposits in reserve), then $R = 10\% = 1/10 = 0.1$. So $m = 1/0.1 = 10$.

This means that, simplistically, if a bank has £100 million in deposits, with a reserve ratio of 10% it can in theory create £1,000 million in loans. If we decreased the reserve ratio to 5% (0.05), then the money multiplier is $1/0.05 = 20$. So that £100 million could create £2,000 million in loans. The lower the reserve ratio, the greater capacity to create money in the economy.

The second type of multiplier is a fiscal multiplier. The word fiscal is here related to the use of government income via taxes and expenditure to influence the economy. Assuming no change in tax rates, if the British government increases spending by £100 million and the UK's GDP increases by £150 million, this is said to have a spending multiplier of 1.5 (£150 million/£100 million). The rationale behind is that if the government spends £100 million on a shiny new hospital, they will employ staff to build and work in this hospital. This extra income to these individuals will then be spent in shops. This in turns creates an increased demand in shops, who will employ more staff and demand more products from suppliers. These suppliers will then have to hire more staff. This rise in employment creates additional income for the economy, and some of this will be spent on retail. There is a resultant increase in demand in the economy and this cycle continues. This is a circular flow of income and spending.

*

CHAPTER 11

Understanding multipliers can give us individuals a chance to try to understand decisions behind central bank rate movements or government expenditure decisions. But how can the machinations of investment bankers and hedge fund managers make an impact on us?

When I realised that being a cosmonaut was perhaps an unrealistic career aspiration, around the age of 15–16, my ambition was to work with numbers in some capacity, perhaps in the financial markets. I started out my Maths degree at Lady Margaret Hall in Oxford, but then transferred to complete a Maths & Economics degree at Royal Holloway, part of the University of London.

I had spent a gap year before university working at KPMG but it was my internship summer in 2006 that sealed my initial career in banking. I had joined Lehman Brothers, the ambitious 150-year-old American investment bank in London in its gleaming tower in Canary Wharf. It took only 25 minutes to get from my front door in East Ham to my seat on the trading floor.

Internships in the bank world are a 10-week beauty parade, though one with minimal sleep. Gathered are some of the sharpest young finance minds from universities in Europe, America and further afield, all of them trying to impress sufficiently to secure that elusive and lucrative graduate job. I had two 5-week stints, one in Equities (event driven and arbitrage trading) and the other in Fixed Income: synthetic CDOs.

I wasn't exactly sure then what CDOs were and perhaps I'm not certain now either, but CDO is an acronym for 'collateralised debt obligation' and possibly these were partially to blame for the global financial mess of 2008. A CDO was a promise to pay an investor a portion of the cash flows from income sources such as mortgage payments (subprime at one stage) or stock dividends. The 2015 film *The Big Short* based on Michael Lewis's 2010 book of the same name, describes the actions of some of the key players involved in the creation of the credit default swap market who sought to bet against the CDO bubble. At its height, there were even CDOs-squared – CDOs backed by other CDOs! With time, we may have even had CDOs-cubed! At one stage, my Lehman team sold CDOs to the Vatican City and I remember seeing the Vatican City headed paper addressed to Lehman, thanking them for this investment opportunity …

When I joined the company properly a year later, I started out as a trader in the European equities markets. I could describe it as being a real-life computer game: several flashing monitors, phones in both hands, shouting across the floor to secure trades. (My younger brother, now a trader himself, assures me that the advance of technology and algorithmic trading have made trading floors much quieter and more cerebral places now.)

My team generated income in 2 ways: gaining commission from facilitating client trades, and then making proprietary income by betting on the markets ourselves. There was a simple mantra: buy low and sell high, or sell high and buy low (the latter known as short-selling, and sometimes portrayed as vulture-like profiting from the carcasses of companies in the doldrums). Ultimately that's all trading was. You can use complex mathematical models to help to understand the market you are involved with, but essentially you want to buy your asset class as cheaply as possible and sell it on for as high a price as possible. For the more complex financial products, particularly in the fixed income world, it was almost like a pass-the-parcel game. You were fine as long as you weren't left holding onto the package when the music stopped.

But the music did stop for the financial markets on 15 September 2008 when Lehman filed for Chapter 11 bankruptcy protection. Some analysts blame this demise on Lehman's involvement in the subprime mortgage crisis and exposure to illiquid assets. This led to a series of unfortunate events where there was a colossal exodus of their clients, tumultuously dizzying drops in their share price and a fatal devaluation of the firm's assets by external credit rating agencies. Ultimately the bank was unable to trade legally with its counterparties.

My peer group, who had joined perhaps expecting a conveyor-belt pathway to riches, had to re-evaluate. A few did stay on in the financial world in areas such as accountancy or insurance; others moved industries altogether, one into fashion, some into setting up their own businesses. One peer, Nikolay Storonsky, now stands on the verge of being a self-made billionaire through the founding of his digital banking start-up Revolut!

After Lehman collapsed, I spent a short time at Japanese bank Nomura, where I learnt that due to an old historical Japanese tradition, junior bankers would often sit nearest to the door in a nod to a historic quirk from Edo era Japan when juniors sat there to prevent any samurai attacks on their bosses. I did eventually qualify as a chartered accountant at PwC, primarily auditing accounts of major listed blue chip companies as well as spending a period of time assessing values of toxic assets that banks had accumulated during the boom days.

Speaking about his *A Brief History of Time*, Stephen Hawking once said, 'Someone told me that each equation I included in the book would halve the sales. I therefore resolved not to have any equations at all.' I appreciate his sentiment and I apologise in advance to my publisher for what looks like potentially a horrible commercial gaffe on my part.

However, there is one mathematical equation that cannot be overlooked. It gets its fair share of criticism for breeding exuberance and overconfidence in the markets. It is the Black–Scholes equation. This and its descendant equations have been blamed for blowing up the financial world.

$$\frac{\partial V}{\partial t} + \frac{1}{2}\sigma^2 S^2 \frac{\partial^2 V}{\partial S^2} + rS\frac{\partial V}{\partial S} - rV = 0$$

OK, I agree that this equation definitely looks icky and nightmare-inducing! But it's more important to understand what the purpose of it was.

Although it was only written down in the early 1970s, the model's story starts back in seventeenth-century Japan. In the Dojima Rice Exchange, futures contracts were written for rice traders. A future is a financial derivative, one of those words that sounds toxic to non-financial ears. A derivative is something that derives its price from the price of another asset. At its most basic, a futures contract states that you agree to buy a certain amount of an asset, in this case rice, after a particular period of time, for a price agreed now.

In the thrilling finale of the 1983 American comedy film *Trading Places* Eddie Murphy's character makes a huge profit selling frozen concentrated orange-juice futures contracts. If you are the owner of a large producer of orange juice, being able to secure oranges at a fixed price for the future gives you some comfort for your planning. You can also secure a future price through options. An option gives you the right (but importantly not the obligation) to buy or sell that asset class (frozen orange juice) at an agreed price at an agreed date. The questions that will run through your mind are how much should you pay for those orange-juice options? What are they worth? This is where the Black–Scholes equation comes into play.

'The problem it's trying to solve is to define the value of the right, but not the obligation, to buy a particular asset at a specified price, within or at the end of a specified time period,' says Myron Scholes. Fischer Black and Myron Scholes introduced a model for pricing financial market options. However, it was

Robert Merton who published the first paper providing a mathematical analysis of this model. Curiously enough, it was Merton himself who coined the term 'Black–Scholes options pricing model'.

That mind-boggling looking equation describes mathematically how the price of a financial derivative changes over time. All sorts of derivative markets were influenced by this formula. To get a sense of what this is, we need to understand what a European call option is (the word 'European' actually has zilch to do with location). The textbook answer to that is that it is 'the right but not obligation to buy an asset at a pre-determined price at a predetermined date'.

Consider a European call option on 1 Facebook share with a strike price of $200 and maturity of 1 year from today. So if I pay to enter into this contract, I will have the right but not the obligation to buy 1 share in Mr Zuckerberg's company at $200 in a year's time. Whether I decide to exercise the right depends on the share price of Facebook in a year's time.

So if the share price is above $200, at $210, I will buy the share from the contract at $200 and sell immediately at $210, locking in a $10 profit. If the share price is below $200, say $190, the contract that says I can buy at $200 is worth nada. I can buy more cheaply at market prices.

Since options give you flexibility, they are of value to investors. The question is, how valuable are they? And how does the value change as the option gets closer to the date when it has to be exercised? This is where Black–Scholes can help value the option. It also takes into account the interest rate, as it accounts for the fact that money today is worth more than money tomorrow.

Solving this equation can give us the value of the price of the option. However, the snag we encounter is that we can't agree on what the value of Facebook's share price will be in a year's time. So while the formula is not bullet-proof, the thinking behind the formula is key. Black, Scholes and Merton do not claim to have any idea about how much the stock price will be next year, as much as I can predict the winner of the 2022 World Cup final (England is always my optimistic go-to for this!). What is fascinating is an assumption they made. They believed that a stock's prices rise and fall in the same unpredictable manner as the movement of dust particles around atoms.

If the movements in stock prices are random, they can then be modelled using some of the same mathematics of modelling random movements. The Black–Scholes uses the volatility of the stock, how much it fluctuates over a particular period. The model assumes the price of heavily traded assets follows a geometric Brownian motion. A stock that is more prone to large upward or downward moves will have a higher option price. (Remarkably one of Einstein's first major contributions to science was his 1905 paper, where he explained Brownian motion using the illustration of pollen grains that were being moved by individual water molecules.)

These derivatives became traded assets in themselves, separate from any owning of the actual asset class. You didn't need to have any interest in ever owning rice or oranges from trading their derivatives. Currently the global derivatives market is estimated at $1.2 quadrillion (that is $1,200,000,000,000,000). Some analysts put this is at 10 times the size of the total global GDP.

Can we blame models like the Black–Scholes for the financial crisis that the world is still emerging from? Yes and no. Financial markets do over-rely on models to help them understand and price markets. But at the same time, greed still cuts through. If a profit is sniffed, individuals can use models to justify their decisions. I was once told as a junior trader, 'Make your decision to buy or sell first, execute it and then come up with a nice story to explain your decision. Don't miss the opportunity faffing about!'

It was this sort of lax attitude that contributed towards the global financial implosion in 2008. Images and videos were streamed across the world of Lehman Brothers' staff leaving with their belongings in cardboard boxes. I had my younger brother bring a shopping trolley so I wouldn't cause damage to my back trudging my items back home.

Maths is a tool for us to understand the world. With the financial world, numbers can be used to trick consumers, but at the same time, if we understand the basic concepts then we are in a position to make a better judgement. I always advise my students, let numbers be your friend and you'll have half a chance of navigating this complex financial world. We don't need to understand the inner workings of complex financial mathematics, but there is a benefit in being able to appreciate its impact on the economic world. And we all live in an economic world.

Marty the Magician's Secret Stock Market Method

Marty the Magician has devised a new secret method of working out which stock will suddenly rocket to success in the financial markets on the London Stock Exchange. He has realised that shares in a company called Virtual Reality Sports are set to double in price every day for the next 10 days and then the share price will go flat indefinitely.

You were sceptical at first, but after 7 days of trading, the shares of Virtual Reality Sports have indeed doubled for 7 straight days. At this stage, you go to the bank and borrow £10,000 to invest in shares.

How much profit will you make at the end of the 10 days?

CHAPTER 12

'Is it Too Good to Be True?'

The Numbers Behind Clever Deceptions

Like many people, I once received an e-mail out of the blue telling me that I was the heir to a $100 million fortune, but couldn't access it because my great uncle passed away tragically before he was able to sign over rights to me. Such a shame. But the good news was that I would be able to unlock this fortune simply by sending $10,000 so that an international lawyer could release the funds.

It was the summer holidays of 2001. I was halfway through my A-level course and I thought, 'Why not have a bit of fun?' I responded to the e-mail, though from a new Hotmail account. 'Alfred Tacker' came into existence that morning. (Apologies to the real Alfred Tack – I was looking around my room for inspiration for a name and found a dusty maroon-coloured book from the 1960s by a writer called Alfred Tack, and just adapted his name.) I decided 'Alfred Tacker' would be a senior figure in my fictional concoction called the

CHAPTER 12

Blue Morning Crystal Believers (BMCB) group. I then proceeded to have an e-mail dialogue with this person who claimed that I was close to receiving my $100 million. Over the course of the summer, I asked my 'friend' to show his allegiance to the BMCB through a series of challenges before we were able to transfer him the $10,000. It culminated in my friend having to send us a photo of himself painted in blue, wearing white trousers and a white hat. He did so, and unwittingly would have made a solid audition for a live-action Smurf. At this stage he grew increasingly frustrated and the e-mail trail withered away as he realised I knew he was attempting to scam us.

Why am I telling you this story? I think part of being numerate is being able to look behind the headlines and examine the figures yourself, and ask if they make sense. Being offered $100 million as a teenager sounded exciting. Nowadays, I have learned to be more sceptical and often subconsciously follow the mantra, 'If it sounds too good to be true, then it probably is.' Mathematics offers us tools and strategies for being able to look at data and information and to make an informed decision ourselves about its validity.

Back in the autumn of 2015, I was at my youngest brother's graduation. The graduation speech was delivered by Mark Damazer, the Master of St Peter's College, Oxford. Damazer, a former controller of BBC Radio 4, was sharing his wisdom about how his newest graduates should embark into the wide world beyond university. He made one statement in particular that has resided with me ever since: while graduates will learn many technical points from their degree, what they need entering the outside world is the guiding mantra that 'scepticism is a virtue, but cynicism is a parasite'. I found this so profound that I had my Year 7 form class write this on the front of their form book and we dissected what this meant.

It is healthy for us as informed individuals to approach what we see around us with a small dose of scepticism, in the sense that we should be willing to question accepted opinions and not just follow them like shuffling sheep. This allows us to question assumptions to see if they make sense. However, cynicism, where we are paranoid about the motivations of everyone, can be an isolated and reclusive place to wind up in. Mathematics can offer us the intellectual tool box from which we can take out the right equipment and methods to delve beneath the immediate facts offered to us.

As an auditor for PwC, my role involved helping to inspect client financial statements. Big accounting firms sign off on the annual set of accounts for corporations saying that the accounts represent a 'true and fair view' of that company's position. It doesn't mean that the accounts are perfect or error free, just that they are free from material misstatements (untrue information in a financial statement that could change the financial decisions of someone who is relying on the statement) and faithfully represent the financial performance and position of that entity. As a junior auditor, one of my roles when examining a client's accounts was to see if anything didn't look right. The vast majority of the time, there is a perfectly sensible explanation, but on very rare occasions, something more untoward might be lurking. This is where maths can step in.

When we think of the evidence left behind from crimes such as fingerprints or strands of hair, we are thinking of physical clues (even if it's biscuit crumbs on the floor near your brother when he claims he didn't eat the last biscuit!). However, there are instances, whether looking at businesses or even science results from a lab, that a fraudster can get ensnared because of their use of numbers. And this is where we meet one of the most counter-intuitive laws, which astonishes all who encounter it for the first time.

If you look at any newspaper, plenty of numbers appear in many contexts such as 'increasing stadium capacity to 62,000', 'celebrating the 50th anniversary', 'cut by 5.8%', 'child aged 12' or 'see inside for 3 tips'. At first glance, all these figures appear completely unrelated to each other, but is there a relationship between them at all? What proportion of numbers would you expect to start with a 2 or a 5 or a 7?

You would expect the first digit to be randomly and evenly spread, in other words it being equally likely that a number you pick would begin with a 1 or a 9. But, if you start going through a magazine, a newspaper or a website and start collecting the data, the following pattern will emerge:

The first number	Percentage probability of beginning with that number
1	30.1%
2	17.6%
3	12.5%
4	9.7%
5	7.9%
6	6.7%
7	5.8%
8	5.1%
9	4.6%

How can a pattern emerge from something as random as numbers plucked from a newspaper? Surely numbers from here are as random as random can be? Well, this strange distribution is determined by something known as Benford's law.

Frank Benford was a research physicist at the industrial behemoth General Electric in the 1930s and he popularised the law. However, it was Simon Newcomb, a Canadian-American mathematician, who first discovered this when he found something bizarre about a book of logarithmic tables. The earlier pages of the book showed more wear and tear than the latter stages, indicating that numbers starting with the digit 1 were being looked up more frequently than numbers starting with 2 through to 9. Based on this initial observation from Newcomb, Benford spent years collecting data to show that the pattern was prevalent in nature. In 1938, he published his results, citing more than 20,000 values such as atomic weights, stock prices, river lengths, numbers in magazine articles and baseball statistics.

Benford's law is now one of the tools used in the accountancy world to detect fraud. In the early 1990s, Dr Mark Nigrini, an Accountancy professor from Dallas, asked his students to use accounts of a business they knew to test Benford's law. One student tested the financial books of his brother-in-law's hardware shop. He discovered that 93% of the numbers began with the digit 1, compared to the expected 30%. The student had stumbled across a fraud!

We have seen Benford's law being applied to other fields too. In 2006, Mebane studied presidential election votes by testing whether the second digits of the reported vote obeyed the frequencies as predicted by Benford's law. In 2007, Dieckmann applied this law to detect falsified scientific data. So if someone is a smart fraudster, they should consider Benford's law and try to make their fabricated data fit this distribution (though not too well!).

So what's the maths underlying Benford's law? A precise mathematical relationship holds: the expected proportion of numbers beginning with the leading digit n is $\log_{10} ((n+1)/n)$. Any calculator can help you to check the results from our earlier table. For $n=1$, so digits starting with the figure 1, we substitute 1 into n to give us $\log(1+1)/1 = \log 2/1 = \log 2 = 0.301$. This is

30.1%. Then try it for n = 2. Log $(2+1)/2 = \log (3/2) = \log$ 1.5 = 0.176 = 17.6%.

Why does this law make sense? Think about a small company with 100 staff. For the company to reach 200 staff and change the first digit of its total employees, the company would have to double in size. This is a 100% increase in company size. However, once the company is at 200 staff, to get the first digit to change now means it has to get to 300. This is only a 50% increase from 200, which should be a little easier than the first 100% increase. Once the company reaches 1,000 staff, we have the same pattern emerging. To reach 2,000 staff is a 100% increase, but then to reach 3,000 is only a 50% increase, and then to reach 4,000 is a 33% increase. So company sizes tend to remain in the low hundreds, thousands or millions. This isn't a perfect mathematical explanation but gives you a sense of why the digit 1 is more common as a starting figure.

So can Benford's law be the panacea for all world problems? Will a new global order of harmony and peace ensue? Should a new religion be formed beatifying Benford? Sadly not. A critique of Benford's law is that data can often diverge from this law for perfectly innocent reasons as suggested earlier. As long as the sample of numbers is sufficiently large and the numbers aren't straitjacketed by some rule, Benford's law seems to work and the first figures of a set of data follow the distribution from the table above. So where would Benford fail? For example, using mobile phone numbers in the UK would not work as they are 11 digits long and all start with 0. Or the height of females in the UK, most of whom will be between 140 and 180 centimetres.

However, accepting these pitfalls, Dr Nigrini said that 'I foresee lots of uses for this stuff, but for me it's just fascinating in itself. For me, Benford is a great hero. His law is not magic, but sometimes it seems like it.' The law stands true regardless of the unit used for measurement. Benford's law has gained an almost cult-like following and has been attributed by some as an underlying property of our universe.

In 2008, British illusionist Derren Brown claimed that he had developed a 'guaranteed' method for winning on the horses. Time and time again, Brown successfully predicted the winner of horse races for a single mother called Khadisha from London, and each time she won by following his advice. In the thrilling conclusion to this series of bets, and as part of a programme called *The System* on UK's Channel 4, the public would find out if the single mother would gamble all her life savings on one final horse.

The answer is yes, she did! But obviously, I'm not about to tell you that Derren Brown could predict the future. During this documentary, he explained how he did it. Jordan Ellenberg, a Mathematics professor from the University of Wisconsin, said that by doing this, Brown probably did 'more for math education in the UK than a dozen sober BBC specials'.

What exactly did happen? In the advertisement for this programme, Brown explained that the 'System is a way of taking a random person and telling them again and again which horse will win. It is 100% guaranteed.'

Let's examine the story in more detail. We followed the journey of this single mum, Khadisha. Initially she received a text message telling her a horse would win a specific race, which the horse proceeded to win. She wasn't told who the text was from but was instructed to start filming herself on camera as she continued to receive texts about winning horses. She started to back the correct horses. Khadisha won in 5 consecutive races with bets increasing in magnitude each time. She visited a racecourse for the first time and, this time, a camera crew was there to film the action. It appeared as though her horse was going to finish well behind the pace but astonishingly the 2 leading horses dismounted their jockeys at the last fence – so she rejoiced in triumph again!

It is at this stage that she was introduced to the Mystic Meg style genius, Derren Brown, who devised 'The System'. On the seventh and final race, Khadisha borrowed £4,000 to place her bet and made it patently clear to the audience that it was money beyond her means. Brown then revealed that all was not as it seemed.

Khadisha was not as special as she first thought. Brown was actually working with 7,776 people $(6 \times 6 \times 6 \times 6 \times 6 = 6^5)$ and Khadisha was the only person who had won all the races. In the first race, they were each given 1 of the 6 horses to back. This left only 1/6th of the original pool of people available, that is 1,296 (6^4). This process was continued until they were only left with 6 individuals, who had been randomly allocated the winning horses in the first 4 races.

This illustrates the number of people still in at the beginning of each race:

Race 1: 7,776
Race 2: 1,296
Race 3: 216
Race 4: 36
Race 5: 6

A camera crew now followed all 6 as they placed their bet for the final race using 'The System'. Of course, 5 of them lost and only 1 remained.

In the final filming with Khadisha, she placed £4,000 on a horse. Of course, there was only a 1 in 6 chance that she would win. Her horse didn't win, leaving Khadisha heartbroken but Brown then revealed he'd put the money on the winning horse without Khadisha realising and hence won £13,000. In reality, he would have bet on every horse in the race to make this outcome guaranteed to occur. (Before you write to Derren Brown showing concern about the ethics of the financial losses incurred by the other 7,775 participants, we were assured that all involved were offered refunds for any cash spent.)

Brown compared his 'System' to homeopathic remedies, explaining that some people feel as though they work (with the placebo effect suggested as one cause) while many others feel as if they don't. Khadisha was adamant that the 'System' was working in her favour, but in clinical trials it didn't (which in this case were the 7,775 other participants). This did seem to make the public genuinely stand up and think about the probability behind the outrageous claims that we sometimes hear in the media. It's all a case of perspective.

I was reading the online *Daily Mail* comments section about another Brown programme where he predicted the lottery numbers, and the comments give us an insight into perspective. One reader, Alistair from Salem, USA, said that he wasn't impressed as he had 'successfully predicted which [lottery] numbers weren't going to be drawn' and that he'd done that 'twice a week for years'. Alistair was successful, but not at his objective of actually winning the lottery.

What happened to Khadisha as part of a TV programme has also been used in the finance industry before – at least apocryphally. The urban myth tells of a stockbroker who sends a letter every week for 10 weeks predicting whether a particular stock will be up or down that week, and predicts this binary outcome perfectly each time. Would you trust this stockbroker?

As humans, we tend to extrapolate results that we see in front us, without necessarily considering what the bigger picture is. The stockbroker sends out 10,240 letters, of which 5,120 says the stock will be up and the other 5,120 says it will be down. If we make the assumption that the chance of a stock being up or down on a particular week is 50%, then he will lose the interest of half of the letter receivers.

As with Khadisha's example, we can trace the success of the stockbroker's letters each week. The number of people who are still involved at the start of each week is as follows:

Week 1: 10,240
Week 2: 5,120
Week 3: 2,560
Week 4: 1,280
Week 5: 640
Week 6: 320
Week 7: 160
Week 8: 80
Week 9: 40
Week 10: 20
Week 11: 10

So by the start of week 11 there will be 10 people who will each have received 10 perfect predictions. And they will think that this stockbroker has an almanac from the future, like the *Grays Sport Almanac 1950–2000* in *Back to the Future Part II* that the character Biff gives his younger self which has the results of all major sports results in the coming 50 years. Of course, we are now wise to the nature of this scam and realise that this stockbroker has anything but a Nostradamus-like ability to predict the future. This scam, as it were, can be applied to anything where there is a limited number of outcomes, such as the football pools where teams can win, lose or draw.

'IS IT TOO GOOD TO BE TRUE?'

With many investment schemes in the finance industry you will read the small print that 'past performance is not indicative of future results'. Just because a fund has 10 consecutive years of success does not guarantee it will continue to outperform in the next year. It may do, of course, but you need to think beyond the headline information you see. Mathematics can offer you a tool to step back and ask: does this make sense? It can't provide you with all the answers, but if it can make you hesitate just for a second and try to grasp what might really be going on before making a decision, then it's a power for good.

Catch Me If You Con : The Game Show with a Panel of 10 Star Hosts

There is a new and unusual TV game show coming soon called *Catch Me If You Con* where contestants will try to trick the show hosts into believing certain things.

The hosts are allowed to ask one question each to the contestants to ascertain if they are telling the truth. A star panel of 10 hosts has been organised for a Comic Relief special including Jeremy Paxman, Jeremy Vine, Sandi Toksvig, Bradley Walsh, Shaun Wallace, Richard Ayoade, Nick Hewer, Rachel Riley, Richard Osman and Alexander Armstrong.

Task 1: Which quiz show is each of the hosts most connected with?
Task 2: In how many different orders could the 10 stars ask their questions?

CHAPTER 13

'The House Always Wins'

The Maths Behind Gambling

December 2014. Six suited and booted British Asians turn up at the glamorous, glitzy and gilded casinos of Caesar's Palace Hotel in Las Vegas. Armed with a hundred dollar bill each, we were ready to take on the mighty house of the casino. Between us, we had professional financial markets traders, tax experts, consultants, maths teachers. A keen sense of numbers coupled with a smooth blend of calculated risk taking and forensic analysis oozed from our group. A crack team assembled to bring home some lucre. Did it happen?

I'd like to say that we won hundreds of thousands and came back to a heroes' welcome at Heathrow airport and I'm now writing from a golden beach on a private Caribbean island while sipping my favourite ice-cold ginger beer. But the reality is not that rosy. Out of the 6 of us, 4 lost our stakes, including myself, and only 2 managed to take home a small profit. The house had beaten us.

However, this was actually not a major disappointment. For those who have seen *The Hangover* comedy film franchise, you will recall that the protagonists stayed at Caesar's Palace. In the first film, the character Alan wins $82,400 playing blackjack by counting cards. But back in real life, the 6 suited characters in my visit were me and my siblings and cousins. We had all accepted that the $100 each could be counted as paying for entertainment, much like we might do for a football match. So the loss was not taken too sorely.

East Ham High Street was once on *Tomorrow's World*, a BBC TV series reporting on new developments in science and technology. Former *Countdown* TV show host Carol Vorderman was there in the mid-1990s, marvelling at some of the first automatic rising bollards in England to slow down traffic. Now in 2018, the very same High Street is peppered with multiple bookmakers, allowing gamblers to take their chances on a multitude of betting opportunities. There are perhaps more bookmakers than any other single type of shop.

Gambling is big business. According to the most recent figures from the Gambling Commission in the UK, the total size of the industry in the UK to the year ended March 2017 was £13.8 billion. The arrival of the internet has transformed opportunities for punters to take a gamble with regular betting, bingo, casinos, lotteries and poker. In excess of 2 million adults in the UK are either problem gamblers or at risk of addiction. Some reports say that there are 24 million active betting accounts in the UK, which is staggering considering the total population is only 66 million.

In moderation, like most activities, gambling can be seen as a fun leisure activity. In excess and out of control, it can pose grave risks to mental health and wealth. So what is the allure of gambling? You only need to flick through any commercial TV broadcaster during advert breaks to see the proliferation of ads tempting you with the latest odds on a football match or the chance to sign up to an introductory offer for an online casino.

Gambling in its various guises encompasses many fields: maths, economics, human psychology and science. Researchers have found gambling a field for experimenting on the world of random events too.

As a fan of sports of all types and a mathematician, gambling is something that has always held a fascination for me. Nowadays this is mostly as an observer, an armchair pundit, pontificating on the rationality of the odds. But for a short while during my early 20s at American investment bank Lehman Brothers and then Japanese bank Nomura, it was almost considered essential training to show flair and pizzazz in finding the right bet.

Part of my role as a financial markets trader involved calculated decision making on investments, which may have seemed like gambling to an untrained outsider. When making investment decisions, particularly for the short term (which can be as short as a few minutes or even seconds), I took speculative positions on various asset classes, usually stocks and shares of listed companies.

To beat the market as a trader or investor you have to find an investment that is mispriced, have some information on the investment that the price doesn't reflect, or be fortunate. In the world of gambling, you need to be able to notice when the odds available are too charitable, have some information that the odds of that event don't reflect, or be fortunate. Looked at from this perspective, both appear very similar but do not be fooled. Despite how it may seem, gambling and investing are not the same.

When you hold a diversified portfolio of stocks, shares and other asset classes, you are highly unlikely to lose everything (unless you happen to be solely buying Lehman Brothers shares on the day they went bust), especially not in a very short time period, such as a day. Despite periods of painful losses in recessionary markets, you will still usually have something. Gambling is not like that for most of us. It is designed for us to lose in the end.

As professional gambler, and *Only Connect* host, Victoria Coren Mitchell has pointed out many times: 'Casino rule 1: the house always wins.' Over short-term periods, some of us can beat the bookies or casinos, but over the long term, only a lucky few or exceptional souls actually pull out a profit.

Gambling is defined as the betting of something of value (the stake usually being money) on an event, or series of events, with an uncertain outcome. The main object of the activity is to secure a prize, usually money or sometimes material goods, for correctly winning the wager. So taking this apart piece by piece, gambling requires 3 separate components to work: the consideration (the value offered and accepted going into this contract); chance (the probability); and the prize (the loot you get to take away). Mathematics offers us insights into how much you should consider staking, the dynamics of the probabilities involved, and the expected return.

So how does the mechanics of sports betting actually work? As a football fan, I will often hear things on the TV such as 'West Ham are 4 to 1 to beat Manchester United' or 'the odds on Iceland beating England at the Euros is a tempting 8 to 1'. For us to understand how bookies stack the odds in their favour, we need to speak their language.

Let's introduce a little bit of algebra, nothing too scary, just 2 little letters in the alphabet, x and y. The odds of 'x to y' stand for a relative probability of $y/(x+y)$. With the use of numbers, this makes more sense. A long shot 9–1 event (9 to 1) is $1/(9+1) = 1/10 = 0.1$ or 10%.

We could also work backwards. Given a relative probability, we are able to calculate what the odds would be. So, if we are given a relative probability of 25%, this represents odds of $(1-0.25)/0.25 = 0.75/0.25 = 3/1$. So a 25% relative probability is 3:1 or 3 to 1 on.

On a Saturday at 3pm, it is a ritual for football fans across the country to go to support their football team. Football is like a religion for the masses. Our cathedral is the stadium and we are the choir, chanting out our songs, with the football players our religious idols. Indeed, my first ever pay packet during my gap year between school and university was spent on a West Ham season ticket to the old Boleyn Ground at Upton Park. Since then, I try to attend home games when I can.

My routine was religious. I left our house in East Ham around 2.40pm, bedecked in a slightly large claret and blue West Ham football shirt, with a scarf draped around my neck. As long as I wasn't running late, I nipped into our local bookie and placed a fiver bet on West Ham to win (I didn't want to be cheering on the opposition!). I justified this almost weekly loss of a fiver as I wouldn't buy the match-day programme (saving £3) and I would bring a sandwich from home instead of buying a Cornish pasty at the club shop (saving a couple of quid at least).

Hearing the growing crescendo of the crowds singing urged me closer to the ground. I'd check my digital watch: 2:57pm – time to make a dash for my seat. Fortunately my seating was in the east stand, fondly called the 'Chicken Run' stand by fans. This was miniscule by modern-day standards, and thus it was easy to navigate through the turnstile and be in my seat in time for my personal highlight of the match: the club song.

The music blared from the stadium loudspeakers and we welcomed the players raucously to the sound of the club's anthem 'I'm Forever Blowing Bubbles'. As long as I arrived with the eternal optimism of my £5 bet slip and was on time for the first verse of our club song, all was right with the world.

In the football match, there are 3 outcomes that can occur: a home win for the mighty Hammers, a draw or, God forbid, a loss for the men in claret and blue. Let's create a hypothetical match in the future: the Champions League semi-final second leg of 2028, where West Ham take on Barcelona in the colossal Camp Nou stadium.

The odds might be a home win for Barcelona at evens (1–1), a draw at 2–1 or an unlikely Hammers smash and grab victory at 5–1. These odds can be converted into relative probabilities:

Barcelona win: 1–1 is a relative probability of 1/2 = 50%
Draw: 2–1 is a relative probability of 1/3 = 33⅓%
West Ham win: 5–1 is a relative probability of 1/6 = 16⅔%

If you add the percentages up, we get a total book of 100%; this represents a fair book.

Let's imagine I own a bookies, imaginatively called Bobby's Bookies, and I allow 3 mates to bet £50, £33.33 and £16.67 against each other to win £100. This is what my theoretical 'book' would like in Microsoft Excel (or any other spreadsheet):

Result	Bet £	For	Against	Relative Probability	Win £
Barcelona win	50.00	1	1	50.00%	100
Draw	33.33	2	1	33.33%	100
West Ham win	16.67	5	1	16.67%	100
Total	100.00			100.00%	

In this particular example, Bobby's Bookies is performing a social function as I stand to make zilch from facilitating this bet. One of my 3 friends will take home the entire £100 gambled between the 3 of them. This is known as a fair book – something you are almost never going to see with a real bookmaker, unless there was an error of some sorts.

If I was looking to make a profit from facilitating these, I would of course reduce the odds. Nothing unethical about it, just see it as a fee for organising this bet. If I wanted to give my 3 pals the same odds relative to each other, we can maintain the same relative probabilities (3:2:1):

Barcelona win: 4–6 is a relative probability of 6/10 = 3/5 = 60%
Draw: 6–4 is a relative probability of 4/10 = 2/5 = 40%
West Ham win: 4–1 is a relative probability of 1/5 = 20%

Now, Bobby's Bookies is in the position where my actual 'book' is greater than 100%. This is known in bookie's lingo as 'overground', 'bookmaker margin' or even the 'vigorish' or 'vig'. These represent the profit that I would expect to make as a bookmaker.

So in an ideal, blue-skies situation, I would accept £120 in bets at my quoted odds in those proportions. Irrespective of the result of the huge Barcelona–West Ham fixture, I would only pay out £100 (including any returned stakes).

What are the different payouts for each of the different possible results?

A £60 bet at 4–6 returns £100 for a Barcelona triumph.
A £40 bet at 6–4 returns £100 for a drawn match.
A £20 bet at 4–1 returns £100 for a (let's be honest) improbable West Ham win.

Result	Bet £	For	Against	Relative Probability	Win £
Barcelona win	60.00	4	6	60.00%	100
Draw	40.00	4	10	40.00%	100
West Ham win	20.00	4	1	20.00%	100
Total	120.00			120.00%	

CHAPTER 13

The total stakes received by Bobby's Bookies are £120 and I would pay out a maximum of £100 regardless of the result. The surplus £20 represents my tidy little profit, which would be a 16⅔% profit on the turnover $(20 \times 100/120)$. And while this is a simplified example, it demonstrates the built-in margin that enables the house (the bookmaker in this case, but also the casino) to win in the long term.

My most fruitful period of gambling included a combination of overconfidence, lucky streaks and the use of 2 betting strategies: accumulator bets and also exploiting minor discrepancies in odds between differing bookies. This was during the summer of 2006, where I had secured an internship at Lehman Brothers. During those heady years of the mid-2000s, the promise of lots of lucre lured in many of the sharpest young minds from university. Nowadays, tech and entrepreneurial start-ups perhaps offer some of the most innovative opportunities for the young and ambitious, but back then it was the trading floors of investment banks and hedge funds. There were even some trader colleagues of mine who had cut their teeth on the trading floor of sports bookies such as Betfair.

During this period, I experimented with multiple sports, concluding with the football World Cup of 2006. You name a sport, and I probably would have been able to make a reasonably informed bet: American football, athletics, basketball, boxing, cricket, horse racing, hockey, ice hockey, rugby, tennis – they were all fair game. One particular type of bet that I enjoyed, despite it failings on several occasions, was the 'acca', meaning an accumulator bet. It is perhaps one of the most exciting and rewarding types of gamble available to a sports fan. Essentially, it brings together multiple bets. For success in an accumulator, each prediction has to be correct.

Let me demonstrate the potential power of an accumulator bet using an example from the last day of the England's football Premier League on 13 May 2018.

These were the odds at one bookmaker for the higher placed team going into the final match winning that particular game. To convert the fractional odds to decimal odds, we consider the odds of 'a/b' as earlier (notice I've switched the x and y from earlier, but this doesn't alter the outcomes). We do $(a+b)/b$. So Burnley, at 2/1 odds, becomes $(2+1)/1 = 3/1 = 3.00$. This means that a £10 bet returns $3.00 \times £10 = £30$. Obviously the profit then is £20, as £10 was your original stake.

Match	Odds	Decimal odds	Return on a £10 bet
Burnley vs Bournemouth	2/1	3.00	30.00
Crystal Palace vs West Brom	4/5	1.80	18.00
Huddersfield vs Arsenal	1/2	1.50	15.00

Match	Odds	Decimal odds	Return on a £10 bet
Liverpool vs Brighton	1/4	1.25	12.50
Manchester United vs Watford	1/2	1.50	15.00
Newcastle vs Chelsea	4/5	1.80	18.00
Southampton vs Manchester City	1/2	1.50	15.00
Swansea vs Stoke	1/1	2.00	20.00
Tottenham vs Leicester	2/5	1.40	14.00
West Ham vs Everton	3/2	2.50	25.00

So if I had placed all these bets individually for £10 each, the outlay would have been £100 as my stake. My potential return was £182.50, the sum of each of the individual returns, giving a profit of £82.50.

To see the power of the accumulator, we would place a single bet on all 10 football matches concurrently going to the higher placed team (unlikely in reality, of course). We multiply all the decimal odds to give us an overall odds of 287.0438. Multiplying this by our £10 stake gives us a jaw-dropping potential return of £2,870.44 (bit of decimal rounding), leaving a profit of £2,860.44.

This is a simple example of why 'acca' bets are especially prominent with punters, albeit perhaps with less than 10 matches predicted. For a reasonably small outlay, a substantial prize lies in wait for the courageous and incredibly lucky. On the flipside, there is no room for error and a single incorrect prediction will wipe out your winnings. The popularity of a bet like this perhaps stems from our belief that fortune will turn our way, and that we'll eventually strike lucky. It's a bit like entering a lottery, a small entry fee with a small likelihood of a gigantic reward ... the eternal optimism of the irrational mind.

Maths can help us to make rational judgements about the odds being offered by a bookmaker. The concept of expected value (EV) is the amount of money a gambler can expect to win or lose if they placed their bet with those particular odds again and again. You can almost consider it as the average you'll win or lose.

EV = (probability of winning × cash amount won for the bet)−(probability of losing × cash amount lost for the bet)

Let's go back to that last day of the Premier League season where all-conquering champions Manchester City took on Southampton. We want to bet £10 on Southampton winning, thus not losing or drawing. The odds at one bookie were:

Manchester City to win: 1/2, which is 1.5
Southampton to win: 5/1
Draw: 4/1

We can calculate the probabilities from the odds, by flipping the fractions upside down (called a reciprocal) and expressing that as a percentage:

Manchester City to win: $1/1.5 = 66.67\%$
Southampton to win: $1/5 = 20.00\%$
Draw: $1/4 = 25.00\%$

The probability of Southampton not winning, using the sum of Manchester triumphing or drawing, is $66.67\% + 25.00\% = 91.67\%$. However, if we use the odds available on Southampton to win, our potential winnings on a tenner are £40 ($5 \times £10 = £50$ less the £10 stake).

Let's go back to calculating the expected value:

$$EV = (0.20 \times £40) - (0.9167 \times £10)$$
$$= 8.00 - 9.167$$
$$= -1.167$$

What this is means is that we'll lose on average £1.167 with the odds available to us. So we can see that the bookies have stacked the odds in their favour.

One of my favourite tricks was to try to find an edge here. Of course, with one bookie you can easily work out what your expected loss would be, as demonstrated above. As a trader, I was able to scour different exchanges to try to find the most advantageous possible price. In the football betting world, there is a cornucopia of betting websites out there that allows you to compare odds between different bookies. With a fairly straightforward use of Microsoft Excel, I could create EV calculations using the best possible odds from different bookies. So one bookie may have given a very tempting offer on Southampton winning, but another bookie gave good odds (relatively speaking) on Manchester City.

With this, there were (exceptionally rare) occasions where I could guarantee a miniscule profit irrespective of the result. This would only happen on matches where the match involved an outrageous long shot, such as when Trinidad & Tobago took on England in the 2006 Germany World Cup. There was a considerable spread of odds available on the Caribbean side as bookmakers could not collectively agree on how to price them up.

Mathematics can offer us insight into how to place our bets and give us an edge. But gambling is a highly risky affair and sometimes it can become hard to apply cold mathematics in the face of dealing with the much more volatile psychology of humans. Pretty much every gambling ad on the TV in the UK now states either in oversized writing or verbally, 'When the fun stops, stop'. This is a relatively new campaign set up by the Senet Group, an independent body created to raise standards in the gambling sector.

They offer the following tips to stay in control of your gambling:

1. Set your limits at the start
2. Only bet what you can afford
3. Never chase your losses
4. Don't bet if you're getting angry
5. Never put betting before your mates

From my short time as a financial markets trader, I can see robust parallels between the mindset of a gambler and the mindset of a trader. As a trader, it required discipline to tell yourself that, if you reached a certain amount of profit, to cash in and vice versa, if your losses were piling up, to call it a day.

It's almost impossible to continually try to beat the financial markets. This is the reason why it is often more shrewd to invest in a tracker fund that traces the movements of the market as a whole, rather than picking individual stocks. Likewise with gambling, you are up against the bookmakers and the casinos. You have to remember that it is their job to make money out of you. The mathematics can help you understand the risks of a particular bet, but you need to make sure you know when it's time to stop.

Personally, gambling in bite-sized portions can be an engaging activity. You may be in the casino and wager a tenner on the ball landing on a red on the roulette wheel. You may be about to watch a football match, and pop into your local bookies and put on a small amount just to add to the excitement of observing. Yes, sometimes you'll win, but the bookies, casinos and the bet enablers are the kingmakers – and in the long run, they hold the cards. Have fun and, to borrow from a catchphrase from *The Hunger Games* films, 'May the odds be ever in your favour!'

Chapter 13 Puzzle

Emmanuel College Porters and Probability

The friendliest porters at Cambridge University belong to Emmanuel College. David, Paul, Monty, Dave, John, Donna, Irene and Daniel have been setting up to welcome the new undergraduate and graduate freshers at the college.

They have printed out the letters spelling WELCOME TO EMMANUEL on individual A4 cards draped over the front of the porter's lodge. There are 7 cards for the word WELCOME, 2 cards for the word TO and 8 cards for the word EMMANUEL.

The pile of A4 cards that spell the word EMMANUEL accidentally drops on the floor! Assuming the porters don't look at the cards, what is the probability that they will pick up those 8 cards in the right order to spell EMMANUEL?

CHAPTER 14

'Fingers on Buzzers, It's Your Starter for 10'

The Strategy of Game Shows

'Well, I will say that you guys, you're very, very clever. And it was a pleasure to watch this match. Thank you very much, Emmanuel, sadly you have to go home now.'

These were the words of comfort from the formidable quiz master Jeremy Paxman on 27 March 2017 to our team from Emmanuel Cambridge at the end of our valiant 170–140 defeat on BBC Two quiz show *University Challenge* to Wolfson Cambridge. It was the closest semi-final in 12 years.

It was a gutting experience for my side to have exited in the semi-final, especially when I had been busy dreaming about how I would lift the trophy if I won the final (hoist over my head like a World Cup trophy, or politely hold to the side like a school certificate?). Note for any of my students reading this: never think too far ahead, try to focus on the task at hand otherwise you may never get to that future task!

Anyway, many readers will first have become aware of me during my exploits on *University Challenge* as a team captain, bringing my Seagull-esque blend of being supportive, positive and knowledgeable. While I'd like to say that I was a fan of the quiz show from childhood, this would be a lie. Of course, I was aware of the programme, a mainstay of national television since 1962 (apart from a 7-year hiatus from 1987). But even though I had flicked past the show while growing up, my entire cumulative watching time was less than an episode of just one viewing of the Australian soap opera *Neighbours*.

Many viewers of *University Challenge* may not have known that I was on another UK quiz show called *The Weakest Link* with host, 'Queen of Mean', Anne Robinson. Not once, but twice. And not in a blaze of glory either. For those new to this show, it starts with 9 contestants and 1 gets democratically voted off each time until just 2 remain in a final head-to-head clash. I was eliminated in the first round, and then invited back for a special 'first round losers' episode. Hoping to make the final, I was unceremoniously dumped out in the second round this time. As a mathematician, I would say that is a 100% improvement in position from round 1 to 2, but it was still an embarrassment. Thankfully this episode was broadcast in the pre-YouTube era, and hence there is no video evidence of my shame!

After *University Challenge*, a composite Emmanuel College Cambridge team took on a panel of 5 expert quizzers on BBC Two's *Eggheads*. We were narrowly pipped by the brainiacs, so I'm still searching for that elusive victory on a quiz show or game show.

In Britain, flick through any TV guide, and there is a ubiquity of game shows and quiz shows on our screens every day of the week. *Countdown*, *Eggheads*, *The Crystal Maze*, *Mastermind*, *Pointless*, *Only Connect*, *The Chase*, the list goes on and on. Some shows you enter purely for the glory of winning, such as *Mastermind*. Others, such as *The Chase*, involve some prize money. And others are entirely physical, such as *Gladiators* (which is now sadly defunct, though I would happily add to the viewing figures if it were to be resuscitated).

What unites most of these shows is that there is an element of strategy involved, from the perspectives of both the contestant and the TV production company. The contestant (and their team in non-individual pursuits) often aims to beat an opponent or maximise their cash prize. For the production company, their challenges are more taxing. There is some fascinating mathematical thought that goes into how much money they should give out on the show and how they should model the behaviour of contestants, while concurrently making sure that the show is a hit among viewers!

Even in a quiz show like *University Challenge*, you should be rewarded purely for knowledge. Surely the team that will win is the one that has collected the most trivia between them? Usually yes, but not quite always. Strategy can play a part. Individuals gain 10 points for their team for getting a starter question correct. This offers their team the chance to gain 15 points on 3 bonus questions, worth

5 points each. However, should you incorrectly interrupt the quiz master during the starter question, you lose 5 points for your team and give the opposition free rein to answer that question without disruption.

We can use basic mathematics to model this situation, but will need to define some parameters. For the sake of this discussion, let's say that interrupting a starter question early and getting it correct has a 60% chance (though of course this will vary depending on the individual quizzer and the strength of the team). This opens up the bonus set of 3 questions. Generally good teams get between half to two-thirds of their overall bonus answers correct, so for ease let's say 50%.

Teams have been known to adopt aggressive buzzer strategies, where they interrupt Paxman if they think they are likely to know the answer or can guess where the question is heading towards. They're aware that an incorrect interruption, a negative known colloquially among quizzers as a 'neg', will lose them 5 points and open up the starter to the opposition.

Some teams collectively decide that they will be aggressive on the buzzer. Looking at it simplistically, if there are 20 starter questions in a match and a team attacks early on every single question, their expected score would have the following calculation:

12 starters correct out of 20 (60%): $(12 \times 10) + 50\%$ of $(12 \times 15$ bonus points available)
$= 120 + 50\%$ of 180
$= 120 + 90 = 210$

However, they would lose 5 points for every 'neg', which is $5 \times 8 = 40$. So $210 - 40 = 170$ points.

With the remaining 8 questions, they assume that their opposition will answer all of them correctly as the pressure is off and they are able to hear the entirety of the question. The reality is that 1 or 2 questions per match are 'dropped', meaning that neither team is able to supply the correct answer.

So the opposition would get:

8 starters correct: $(8 \times 10) + 50\%$ of $(8 \times 15$ bonus points available)
$= 80 + 50\%$ of 120
$= 80 + 60 = 140$

The more forward buzzing team scores 170 against the 140 of the more passive team. I understand that this is very much a back-of-the-envelope calculation, but this raises a moot point about strategy. We can play about with the assumptions about percentages of starter conversion and bonus conversion. But the main point is that an aggressive team, which is confident that they can maintain their strategy throughout, should be able to outperform a more passive team. The risk-to-reward balance generally favours the braver team on the buzzer. This is largely because a starter is not just worth 10 points, but offers up the bonus sets as well.

In football, a method of pressing attacking football that has gained traction is the 'gegenpress', which Liverpool Football Club has deployed to many plaudits in their charge to the final of the Champions League of 2017–18 under German manager Jürgen Klopp. This technique, originating in the 1970s with the Dutch national team, involves trying to win the ball back immediately after your team has just lost it. So to transfer that over to *University Challenge* style quizzing, you need to keep on buzzing even after a few 'negs'. A high risk strategy, but perhaps my so-called 'gegenpress' of quizzing, where teams aggressively press the buzzer, may become a hallmark of strategically aggressive quiz buzzing teams!

The spring of 2018 saw the return of the quiz show *Who Wants to Be a Millionaire?*, perhaps one of the most successful British quiz shows of all time (even inspiring the book that spawned the Oscar-winning film *Slumdog Millionaire*). I actually used the enticement of a million from the show in an assembly to my school students. But instead of a million *pounds* from a quiz show, I told students about the million *dollar* prize of the Millennium Prize Problems, set up by the Clay Mathematics Institute in 2000, to tackle 7 of the most intractable problems in maths. Only one has been solved and a million dollars is the reward for those that can solve another.

But *Who Wants to Be a Millionaire?* offers us some interesting mathematical based choices as well. In the original run of the show from 1998 to 2007 in the UK, the format was as follows:

Every question offered 4 multiple choice options, logically labelled A, B, C and D. Three lifelines were available to aid contestants.

Lifeline 1: 50/50 eliminates 2 of the wrong answers.

Lifeline 2: Phone a friend allows you to ring your best pub-quiz mate for 30 seconds.

Lifeline 3: Ask the audience gives you the opportunity to see how the audience would have answered.

In terms of prize money, contestants answered 5 questions to reach the first safety net of £1,000. After that, they were in theory 10 questions away from a million pounds, with another safety net at £32,000. The safety net meant that if the contestant incorrectly answered a question after that level, they did not fall beyond that particular amount.

£1,000	Safety net 1
£2,000	
£4,000	
£8,000	
£16,000	
£32,000	Safety net 2
£64,000	
£125,000	
£250,000	
£500,000	
£1,000,000	

Eagle-eyed readers will notice that after £64,000, the prize fund does not quite double each time. If it did, it would be £128,000; £256,000; £512,000; and then finally

£1,024,000. These are all powers of 2 or, as mathematicians write it, 2 to the power of n (2^n). It wouldn't quite have the same ring to it – *Who Wants to Be a Millionaire and 24 Grand?*!

While maths can offer us insights into the strategies that can be employed playing this quiz show, it can offer us a perspective from which to view how different types of contestants play the game. The final question of the game sees contestants sitting pretty on half a million. Just a question away from securing a million pounds. The question at this stage is: how much are you willing to risk losing £468,000 (the half a million less the £32,000 safety net) to gain an additional £500,000. It depends ultimately on how much you value the £1 million.

The 'dismal science' of economics, a term coined by Victorian historian Thomas Carlyle, can aid us now. The concept of utility comes to play – how much use does a consumer gain from a good, in this case the money or £1 million. To the majority of us mortal folk, £32,000 is a large amount of money and would make a significant difference to our lives, at least in the short term. However, to a wealthy person, another £32,000 of course is simply an additional sum – the utility is relatively low. If you were to offer them a million pounds, then I'm fairly certain they'd think the utility for them was worth chatting about.

Picture a scenario with 3 different contestants:

Xavier: Someone who borrowed heavily during the easy credit days before the 2008 financial recession. Even £4,000 or £8,000 would make a significant difference to his whole outlook on life.

Yvonne: A successful accountant in her late-40s. Doesn't need the cash desperately but £64,000 would help pay off her mortgage.

Zara: An incredibly successful fashion designer who lives a luxury lifestyle. But a £1 million payoff would still enable her to buy a Bugatti Veyron sports car.

For Xavier, if he reaches the safety net of £32,000, the value to him starts becoming irrelevant as the small amounts have already changed his life. For Yvonne, amounts up to £32,000 would not make much of a difference, but once we start entering into the territory of £64,000, this becomes serious money. And for Zara, she is moderately indifferent at the lowest levels, but perhaps at £250,000 she will start thinking about what she might be able to buy with this amount. So when looking at contestants, we should try to evaluate what utility levels they place on the cash.

We can actually use a mathematical technique called stochastic dynamic programming to help us make decisions at specific junctures in this quiz game. This technique was introduced by Richard Bellman in 1958 to help us model solving problems of decision-making under uncertainty.

Let's examine this with some scenarios:

'FINGERS ON BUZZERS, IT'S YOUR STARTER FOR 10'

Scenario 1

You are on £500,000 and you know that 2 of the 4 options are incorrect, say A and B for sure. But you don't know between C and D. You can stop and take home £500,000. Or you can guess and take a gamble on 2 equally likely choices. So you can calculate an expected reward as:

$$0.5 \times £32k + 0.5 \times £1m = £16k + £500k = £516k$$

£516k is your expected 'average' reward if you gamble.

Scenario 2

You are on £500,000 and do not have an inkling about the answer. Using your 50–50 lifeline, you will either get the £1 million or drop back to £32,000. So your expected outcome is:

$$0.5 \times £32k + 0.5 \times £1m = £516k \quad \text{(which is the}$$
same as Scenario 1)

Scenario 3

You are on £250,000 and have used up all your lifelines. You think that the chance of A being correct is 0.6. Also, you assess the probability of B being right as 0.3 and C being right is 0.1, but that D is definitely a no-no and thus a 0 probability. In this case, you can walk away with

£250,000 or you could gamble. So your expected reward is at least:

$$0.5 \times £32k + 0.6 \times £500k = £16k + £300k = £316k$$

In this scenario, we say at least £316,000, because if you get the £500,000 question correct, you have the option to try out the £1 million question.

One game show that I haven't been on but perhaps would like to try one day is *Pointless*. The show's co-host, Richard Osman, once tweeted 'Seagull doing what seagulls do here' after one of my *University Challenge* matches, so the least I owe him is an appearance one day! This show has graced our TV screens in Britain since 2009. In each episode, the quiz features 2 contestants on a team trying to figure out correct but obscure answers to general knowledge questions with the objective of scoring as few points as possible. The questions are factual and the production company have got answers from a panel of 100 individuals before the show. The dream is that a contestant picks a correct answer but one that has not been selected by a single member of that panel – that elusive 'pointless' answer.

The show rewards knowledge of the obscure, trivia at its most extreme. All of us can name a few Steven Spielberg films but what answer might we give if we are trying to think of a film that others won't think of? So although choices such as *ET* or *Jaws* would be very popular, I would back myself in thinking that not many would think of *The Sugarland Express* or the film *1941*. These would gather very few votes and perhaps even 0, gaining that coveted 'pointless' answer.

Four pairs start the show, and over the course of the programme, we wave goodbye to 3 of the pairs. In earlier series, for the final round, the last pair select a topic from a few options presented to them. In order to win the jackpot, they are allowed to give 3 answers. A final round from 2012 involved naming any male England footballer who has scored in a World Cup (non-qualifying match). If any of the individual responses is a pointless answer, that is success for the team and money coming their way! What often happens is that they choose a correct answer, but a few of the panel (perhaps 1 to 5 others) selected that answer as well.

Again maths can step in to help us analyse this game show. Let $x > 0$ be the number of those surveyed among the 100 who give a particular answer. So $x/100$ is an estimate of the probability that a randomly chosen panellist will give that answer. This means that the chance of the person not giving that particular answer is just $1-x/100$. We can continue with this logic: if we have another set of 100 random people, the chance 0 of them gave the same answer is: $(1-x/100)^{100}$

Here's where a little more advanced maths can assist us. A useful approximation for this when x is small compared to N is: $(1-x/N)^N$ is approximately e^{-x}

The letter e, you'll remember from Chapter 10, is a mathematical constant, which is approximately 2.71828 (though it goes on forever) and is the basis of the natural logarithm.

So $e^{-1} = 0.37$, $e^{-2} = 0.135$, $e^{-3} = 0.05$

Even if just 2 panellists in that group of 100 gave the answer, the probability of being pointless with another panel of 100 is fairly small, about 13.5% (derived from the 0.135).

The worst case scenario for a contestant is if they gave 3 different answers, each with 1 person from the pool selecting it. Each answer would have about a 37% probability of being pointless. So we calculate the chance of 0 of the 3 answers being pointless as $(1-0.37)^3 = 0.25$.

The chance that at least 1 of them is pointless is 1 minus this answer, which is $1-0.25 = 0.75$ or about 75%. And this would be the most rotten luck for the contestants in the final round – to have selected 3 options, all with just 1 person from a panel having chosen it.

The next time you switch on your TV and sit back to watch a game show or quiz, please do enjoy the drama and tension from the comfort of your sofa. Shout out the answers as loudly as possible! But do consider the strategies that contestants have to employ when money might be on the line.

Pressure can also make us do strange things. Under training circumstances, taking a penalty kick in football is a reasonably straightforward task for a professional. But under the lights and immense pressure, legs can turn to jelly. Game shows are no different: pressure can make embarrassingly straightforward bits of trivia questions feel like supercharged quantum chromodynamics.

Post-match *University Challenge* Handshakes

In a *University Challenge* quiz match, there are 2 teams of 4 quizzers. After my semi-final match, all 8 contestants shook hands with every other semi-finalist from my Emmanuel College Cambridge team and the Wolfson College Cambridge team. Every single contestant also shook hands with quiz master Jeremy Paxman. So your Starter for 10 puzzle is 'How many different handshakes are there in total?'

CHAPTER 15

'I Get By with a Little Help from My Friends'

The Metrics of Mates

Facebook seems to tell me that I have 3,000 friends. That's one heck of an invite list for my next birthday party! Sure, I'd like to think I'm an amenable guy who can make human connections swiftly, but do I really have 3,000 people that I could pick up the phone to in times of dire need (such as when West Ham crash to a heavy home defeat to Tottenham)? Or is 3,000 just the number of people I have acquainted myself with at some stage in my life?

The same qualities that have allowed me to make friends with all sorts of people, from a broad variety of backgrounds, inexplicably also seem to have kept me single for the past few years as well. My broad ranging interests and hobbies have allowed new friendships to blossom quickly and yet I've been unlucky in love. Can mathematics help understand the nature of friendships and relationships? While these are both human undertakings, the harsh reality of numbers can sometimes help to us to gain some comprehension. In this chapter, we will explore some of the numbers behind friendships in particular.

If you're anything like me, you might have some friends you hit the gym with, some friends you meet up with to gossip, others you call up for a big night out, and perhaps a different set you see to do cultural things such as visiting a gallery or a museum. Horses for courses, as they may say. But have you ever really thought about how the mates in your life influence your behaviour in your life outside of that friendship? Yes, you might know what a particular friend brings to the table when sight-seeing or choosing the best glass of white wine on the menu. But outside of those direct interactions, do they leave their mark?

I have read research that suggests that the calibre of our nearest friends can be an accurate predictor of our own achievements. I have personally always believed that, as individuals, we become the mathematical mean of our closest friends and family. A non-mathematical way of looking at this is to say that we are heavily influenced by those who are closest to us. I am far from perfect, but I think there are some aspects of my personality – my positive outlook on life, resilience to adverse events and even openness to all sorts of experiences – that have been directly influenced and developed from sharing experiences with my inner circle of companions. We can't choose our family, but we can definitely choose our friends. And in the internet age, it's easier than ever to find people with whom you can really connect.

A few years ago, I became aware of Jim Rohn, the American author of *My Philosophy for Successful Living* and a motivational speaker. He believed that individuals are the 'average of the 5 people we spend the most time with'. Following his philosophy, we can apply the law of averages, which is the theory that the result of any situation is the arithmetic average of all the different outcomes.

So without naming the individuals in my life (as they might be upset if they score lower than others!), let me show you what I mean. Let's say I have 5 closest friends with differing levels of 'positivity' – 50, 70, 70, 70 and 90. The arithmetic average of this is a straightforward calculation of summing up the numbers and dividing by 5, so $350/5 = 70$. So in the status quo, my 'positivity' would be 70. If in theory, I was willing to say, 'Oh that friend who is a 50 is drifting away from my inner circle and I could replace them by someone else who has a "positivity" of 60,' then my average would change to 72. So in theory, I've increased my own positivity by replacing one of the inner circle of my friends. The maths is quite straightforward, but the reality of the world is not quite so cynical – and of course there is a lot to friendships beyond positivity!

Picking friends is not like choosing the paint colour on our walls or a new design of teacup, it's obviously a much more complex and personal process. All of our relationships represent an abundance of various

connections. We have friends that we met during childhood, others from our time at school, some from university, others from our work, some from a club or activity
or others through serendipitous meetings. I once made
a friend by repeatedly trading places with a competing
runner during a 10-kilometre race!

The friends that we have had for a prolonged period
of time can often be ones that make the most considerable impact, particularly as we will have shared memo-
ries and experiences that perhaps have shaped the way
we think of ourselves. I'm certainly not suggesting
that we should shed important friendships just to try
to improve a positivity average. But the idea of cumulative influence is worth mulling over, and I think it
does make one consider the maths behind something
that most of us intuitively understand. If a friendship is
totally draining and adds little to you over a protracted
period of time, you may need to wonder what you can
do to resurrect the friendship or even if that relationship has run its course.

Here's an exercise you could try out if you want to
entertain the cold-hearted utilitarian in you: make a
list of the most important attributes needed to help an
individual flourish. For the purposes of this discussion,
let me use loyalty, humour, perseverance, generosity
and positivity. Now, write down what you think your
personal rating between 1–10 is for each of these traits,
with 1 being the lowest and 10 being the highest. Now
write the names of the 5 people you spend most of your
time with (if that means your work boss, then so be it,

he or she has to get included). Assign a numerical value to each person for each category. Then, average out a category such as humour and see whether your level of humour matches the people you spend the most time with. Also, you can work out an average for the overall individual as well.

Of course, the real world has far more shades of grey than this, as friendships are not just about what we can get from others, but what we can contribute to those closest to us! The greatest joys in life are when we contribute to the success and happiness of others. But this exercise is just a perspective where numbers may help.

I know from personal experience that the older I get, the more I admire friends who are thoughtful and energetic. By being around them, it makes me raise my game. If there is a particular trait that is important to your personal being or that you want to encourage in yourself, then surrounding yourselves with people like that can raise your average in that field. If a friend is particularly negative over a lengthy period of time, I naturally find that I will drift from them, even if not deliberately. Perhaps there's a subconscious averaging going on leading to a decision that my own outlook is being brought down. This method of averaging is clearly not a science, especially as there may be more or less than 5 people in our inner circle. Further, assigning each person an equal 20% value of influence may not be true, as certain companions or colleagues have a more significant impact than that. Nonetheless, this is a useful framework for one to think about the impact that individuals have on us.

Oxford anthropologist and evolutionary psychologist Robin Dunbar has made contributions toward the fields of looking at friendships from a quantitative perspective through the eponymous Dunbar number. At the time he developed his theory in the early 1990s, he was at University College London trying to understand why primates devote so much time and effort to grooming. During his investigation, he uncovered a correlation between primate brain size and average social group size. He found that as primates have large brains, they live in socially complicated societies. In theory, you could predict the group size for an animal based on the size of its neocortex and, in particular, its frontal lobe.

Dunbar found that human social groups are structured in a series of layers, almost like an onion, and these layers have very specific relationships to each other. He hypothesised that there is a maximum number of people we can accommodate in each level of friendship, and that humans will have 1 or 2 special friends (perhaps a partner), 5 intimate friends, then 15 best friends, then 50 good friends, 150 'just' friends and finally about 500 acquaintances. Our relationships form a series of concentric circles of increasing size, but they are coupled with decreasing intensity and thus the quality of the relationship decreases too. Roughly speaking, the rule of 3 comes when moving from one circle to another: $5 \times 3 = 15$, $15 \times 3 = 45$ (rounded to 50), $50 \times 3 = 150$ and $150 \times 3 = 450$ (rounded to 500).

The 150 number is key here. By using the average human brain size and extrapolating from the results of primates, Dunbar proposed that we can only comfortably tend to 150 stable relationships. He explained it by saying that this is 'the number of people you would not feel embarrassed about joining uninvited for a drink if you happened to bump into them in a bar'.

The magic number of 150 has cropped up elsewhere in society. W.L. Gore and Associates (a company known for their Gore-Tex waterproofs brand) found by trial and error that if more than 150 staff were working together in the same building, social problems started to emerge. So they responded by having limits of buildings with 150 employees and 150 parking spaces. The Swedish tax authorities even acknowledged the power of Dunbar's research by undergoing a reorganisation that involved an upper limit of 150 per office.

The circles of these structures appear within our modern social networks. Stephen Wolfram, founder and CEO of Wolfram Research, studied a million Facebook accounts and found that most people had between 150 and 250 friends. (It is of course easier to add a 'friend' on Facebook than to actually make a new friend in real life, so it's not surprising that the Facebook friend count is a little higher than Dunbar's number.) The structural organisation of modern armies seem to fit within Dunbar's parameters too. The company size in the British army is 120 troops, with the equivalent across the pond in America being 180. Even the level of the 'intimate friends' circle is reflected in these military examples: British SAS (Special Air Service) patrols have 4 men each.

Dunbar's research was conducted in the early 1990s, before social media technologies really took root. Nowadays, people develop many friendships that exist entirely online, so these rough guide numbers may change for younger generations of 'digital natives'. We'll have to wait to see how the next generation, the so-called Generation Z (born mid-1990s to early 2000s), go through adulthood before we can assess the impact of technology on their friendships.

Jeffrey Hall, an associate professor of Communication Studies at the University of Kansas, calls these different circles of friendships 'close friends', 'friends', 'casual friends' and then 'acquaintances'. Hall did research on how much time you had to spend with someone at different levels of friendship. He found that it only takes a mere 90 hours for a stranger to become your friend, and then a further 110 hours to step into your close friends' circle.

You would think that friendships emerge naturally, and that they can't just be pinned down by a metric as simple as the amount of time we spend with someone. And yet, the research conducted by Hall on adults who had moved home in the past 6 months and were looking for new friends suggests time spent together is a crucial factor.

Hall and his associates found that it takes about 50 hours of spending time together for 2 people to become 'casual friends' and then a further 150 hours (a total of

200 hours) of quality time to establish a close friendship. In my personal experience, particular circumstances and the intensity of the time together can make the transitions between the various circles of friendship happen far faster than usual! Hence why many people form long-lasting relationships at university, where they spend short and meaningful bursts of time with their peers. Or even on the reality dating show *Love Island*, contestants spend every waking moment for several weeks together. This intensity accelerates the depth of their friendships and relationships, as ably demonstrated by the 2017 bromance couple (and my favourites), Chris Hughes and Kem Cetinay, and 2018 runaway winners Dani Dyer and Jack Fincham.

In the final few months of 1999, there were growing concerns and even minor panic that computers would not be able to cope with the calendars turning from 1999 to 2000. The world did, however, transition relatively smoothly to the year 2000. East London awoke to the news on 1 January 2000 that Michael Wilshaw, the formidable head teacher of my Newham secondary school, St Bonaventure's, had been deservedly knighted, for his services to education, in the 2000 New Year Honours. Sir Michael subsequently went on to become the Chief Inspector of Schools in England and head of the inspection service Ofsted from 2012 to 2016. Given the head's knighthood, my fellow pupils were unsurprised to hear that the current Education Secretary, David Blunkett, would be visiting our school on 21 January 2000. As a

Year 11 student and Deputy Head Prefect, I would be one of the students greeting Mr Blunkett.

With a brand new white shirt and freshly laundered school blazer, I arrived at school that morning only to find that the announcement of David Blunkett's visit was a security ruse – the real visitor was going to be Prime Minister Tony Blair. I thought to myself this meant that I was only a step away from being connected to figures such as Kofi Annan (Secretary-General of the UN from 1997 to 2006) or Bill Clinton (US President from 1993 to 2001).

So if I was now perhaps one connection away from these political figures, how many connections did I have to other famous people? Let's say to Michael Jackson. Bill Clinton surely knew someone who knew MJ. So he might have been 3 steps away: Bobby to Tony Blair to Bill Clinton to Michael Jackson. The popstar and former Take That member Robbie Williams was pretty huge in the UK in the late 1990s and early 2000s (who doesn't love belting out his song 'Angels'?). How many steps was I away from him? Kwasi Danquah, better known as the rapper Tinchy Stryder, was at St Bon's 3 school years below me, so Robbie couldn't have been more than 4 steps apart. What about the then-England football manager Kevin Keegan? Former England international Jermaine Defoe was a year above me at St Bon's till he moved to the FA's National School of Excellence at Lilleshall Hall. So surely I was only maybe 3 steps away from Kevin too?

If I picked a random maths teacher from a school in Lima in Peru, how many steps of connection would I be away from this person? How many steps are we, in theory, from any other person on the planet, currently 7.6 billion and rising? Mathematics and graph theory can help turn this question – which sounds impossible to answer at first – into something we can start to calculate.

A theory called '6 degrees of separation' can shed some light here. This phrase gained traction as the title of a 1993 American comedy-drama film adapted from a John Guare Pulitzer Prize-nominated play of the same name. In the film, Will Smith's character dupes elite families into believing that he is the son of actor Sidney Poitier by claiming to know friends of friends.

The original theory was expounded by Hungarian author, Frigyes Karinthy, in his 1929 short story 'Chains'. The concept is that all people in the world are 6 or fewer steps from each other, so that a chain of a 'friend of a friend' can connect any 2 people scattered across the globe in no more than 6 steps. In John Guare's play, the character Ouisa Kittredge says:

Six degrees of separation. Between us and everybody else on the planet. The President of the United States. A gondolier in Venice. Fill in the names. I find that a) tremendously comforting that we're so close and b) like Chinese water torture that we're so close. Because you have to find the right 6 people to make the connection. It's not just big names. It's anyone. A native in a rain forest. A Tierra del Fuegan. An Eskimo. I am bound to everyone on the planet by a trail of 6 people.

It is quite a profound statement that hints at how connected we truly are to each other on the planet. Mathematically, we are 1 degree of separation from any person we know. And then 2 degrees from someone they know (that we don't). While it appears likely that there may some remote tribe in a rainforest that has chosen to keep itself from the world, does the 6-degrees theory appear mathematically plausible?

Let's define a graph as a set of vertices (circles) in which some pairs of vertices are connected by edges (or lines). These lines can be used to represent the relationships between the objects, which are our circles. If we drew a graph in which the circles represented people, the lines between any 2 people mean that they know each other. The distance between any 2 vertices, say X and Y, is the least number of edges that we have to cross to get from X to Y in the graph.

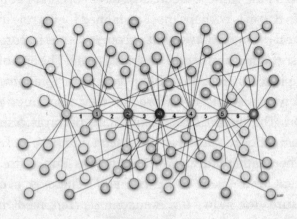

Wikimedia Commons

Let's say that you are acquainted with 300 people. And we'll assume that each of these 300 individuals knows 300 people as well. This gives us $300 \times 300 = 90,000$ people that we might know in 2 steps. If each of these 90,000 people knows 300 people, then this gives us 27 million people within 3 steps. You might be thinking, 'Hey, but surely some of these 27 million people have overlaps?' Yes that's right, but using this simplification helps us to see the big picture.

What we have seen is an example of exponential growth. More simply, if we say we are trying to calculate 'x', the total number of people involved and 'n' the degrees of separation. So $x = 300^n$. If $n = 6$, then we have $x = 300^6 = 7.29 \times 10^{14}$. The world's current population is about 7.6 billion people, that is 7.6×10^9. So while this model ignores overlapping friendships, it does suggest that 6 degrees of separation is most certainly possible.

You may have come across an application of this theory in the game '6 Degrees of Kevin Bacon'. The actor Kevin Bacon is perhaps most acclaimed for his role in the musical drama *Footloose*, but of course he has featured in a host of other films and TV series as well. The aim of the game is to connect any Hollywood actor to another in the fewest possible steps based on their association with Bacon. The measure of proximity to Bacon is actually mathematically known as the 'Bacon number'.

The mathematical community has run with this concept and taken it further still. There is even a measure for how far someone is removed from co-writing a mathematical paper with Paul Erdős, a prolific Hungarian mathematician. So an Erdős number of 0 is reserved for the man Paul Erdős himself. The number 1 would be for someone who co-wrote a paper with Erdős, and then 2 would be for someone who co-wrote a paper with someone who co-wrote a paper with Erdős and so on.

Mathematicians have even gone as far as combining the 2 spheres of Hollywood and mathematics together and concocted the Erdős–Bacon number. Paul Erdős has an Erdős–Bacon number of 3, as he has a Bacon number of 3 and an Erdős number of 0. The late Professor Stephen Hawking had an Erdős–Bacon number of 6. His Bacon number was 2, having appeared alongside John Cleese in *Monty Python Live (Mostly)* and Cleese having acted alongside Bacon in *The Big Picture*. Hawking's Erdős number was 4.

I haven't quite worked out my Erdős–Bacon number yet, though hopefully I'll have a number for you in a few years. I'm currently researching for my doctorate into mathematics anxiety. While this is undoubtedly rooted in education rather than maths, I am hoping to pick up some Erdős points eventually by co-authoring some research papers with someone who has mathematical links at Cambridge. As for my Bacon component, I was thrilled to have taken part in a Cambridge student comedy-improv game show called *University Challenged*, where there were a few aspiring Footlights students. Footlights is the amateur theatrical club in

Cambridge, whose alumni thespians include Hugh Laurie, Clive James and David Mitchell. I'm hoping that via this Footlights connection, my Bacon number will get lower in the future. So check back on my Erdős–Bacon number in a few years from now!

Mathematicians Duncan Watt and Steven Strogatz showed that in a random network, we are able to mathematically quantify the average path length between 2 nodes as log N/log K. A log (short for logarithm) answers the question of how many of one number do we multiply to get another number. For example, $\log_2(8) = 3$ means that we have multiply 2 by itself 3 times to get 8. This is where N is the total number of nodes and K is the number of acquaintances per node. So if we say N = 300 million (90% of the American population) and K = 30, then the degrees of separation = 19.5/3.4 = 5.7.

For those of you who have a Facebook account, you'll be delighted to know that Facebook has shown that the average distance between users has shrunk over time, from 5.28 in 2008 to 4.74 in 2011 to 4.57 in 2016 among its 2 billion or so users. So we are getting closer and closer to each other, online at least – though many people report feeling increasingly distanced from genuine human contact, one of the paradoxes of social media connectivity.

'With a Little Help from My Friends' is a feelgood song from the Beatles' 1967 album *Sgt Pepper's Lonely Hearts Club Band*. The song opens with the lyrics 'What would you think if I sang out of tune? Would you stand up and walk out on me?' If we followed the brutal-hearted arithmetic average that Jim Rohn suggested, yes we might do! But human friendships are of course much more intricate (and messier!) than that. The world of numbers can help shine light on our friendships and social circles, whether they be with good friends, casual acquaintances or bosom buddies. But ultimately, any endeavour involving human emotions and thoughts can sometimes defy logic. And perhaps therein lies the beauty of friendships.

Chapter 15 Puzzle

Friends Party at Tate Modern

I'm organising an exclusive party for my friends at Tate Modern on a Friday night. I set up a Spotify music playlist with the specific amount of songs from my favourite artists.

- 3 × Ellie Goulding
- 1 × Big Shaq
- 4 × Kerry Andrew
- 1 × Childish Gambino
- 5 × Stormzy
- 9 × Jamie Cullum
- 2 × Muse

I want my party to end on some of my most prized upbeat songs so that my guests leave on a feel-good feeling, and I pick the music of Norwegian star Sigrid. I could have chosen any musician to complete this playlist, but I have to choose 6 songs in particular. Why?

CHAPTER 16

'I Should Be So Lucky in Love'

Making the Maths of Love Work for You

Into watching the footy? Can you tolerate support of West Ham? Relax by listening to music? Opera by Verdi, chart pop of Justin Bieber and some rap of Stormzy alright for you? Happy with a night in front of the telly with some comedy of British sketch shows *Monty Python* or a BBC Four documentary by Dr Janina Ramirez on the Vikings? Enjoy wandering around an art gallery searching for nineteenth-century Pre-Raphaelites paintings or zany modern works? Love the rush of a high intensity circuits class followed by some tennis?

If the answer is yes to several of these questions, please send your details to my publisher and we'll set up a date with me!

Thank you to the evergreen Aussie popstar Kylie Minogue for her 1987 international hit 'I Should Be So Lucky'. In the song, Kylie is hoping that a positive karma will propel her towards love. But I want to explore whether mathematics can help me, or indeed anyone, in their quest for 'the one'? This cheery song was written and produced by the trio Stock Aitken Waterman. As a great quiz factoid, Waterman stated that this song was inspired by Johann Pachelbel's 'Canon in D', a classical piece from the late seventeenth century that was voted seventeenth best piece of all time in the Classic FM 2018 survey of listeners.

'When are you getting married, monay?' The word 'monay' sounds like the pronunciation of the French impressionist painter Monet, whose print of the Palace of Westminster in London adorned my bedroom wall for more than a decade. But 'monay' is the word for 'son' in Malayalam, a language from the state of Kerala in south-west India, and is a term of address used by anyone old enough to be your parent in my community, including various uncles and aunties. This question is often fired at me at family and community functions and my response tends to defer a decision to the future: 'I will get married soon, but have to focus for now on my studies' (choose whichever qualification I am currently on). I'm not sure how much longer I can insert appropriate qualifications to delay my response, though my part-time doctorate will extend for a few more years.

While many of my family and their extended community have been settled in the UK since the late 1970s or early 1980s, they still carry the assumption that after one has graduated from university and found a settled job, it is time to get married. Many people still think that arranged marriages dominate in India.

Based on my own experience, from my community of Malayalees (Keralites from India), arranged marriages do exist but 'love' (said in your best Indian accent) marriages are on the increase. With the growth of cities, and young Indians moving away from the ancestral villages in search of education, jobs and opportunities, the youth of India are increasingly intermingling and hence more likely to meet a suitable partner from outside of their historical circles through their travels.

Here in England, my friends would consider anything but a 'love' marriage absurd (and perhaps rightly so), but it is a relatively recent concept historically. There are many people in Britain who have met their spouses through a matchmaker or an interested family member.

As far back as biblical times, arranged marriages existed. Traditionally, marriages were seen as alliances of political, military or social leaning. I try to visit relatives in India every few years, and often once initial pleasantries are exchanged, a relative will usually pipe up with 'Bobby, when are you getting an alliance?' Old terminology takes time to disappear.

I've got to the stage of my life in my early to mid-30s where my Facebook wall is an incessant flood of acquaintances' photos of weddings and babies. A few years back, friends would have been sheepish to acknowledge that they met online. Nowadays, having met online is commonplace, and perhaps shows a sense of taking matters into one's own digital hands, rather than waiting for a serendipitous meeting at a pub or an introduction from a friend. But personal introductions are still a safe way to meet potential dates. If an introduction is good enough for Prince Harry and Meghan Markle, then it should be good enough for the rest of us mortals! (They were apparently set up on a blind date by their mutual pal Misha Nonoo, a fashion designer – and hence they were only 2 degrees of separation from knowing each other.) Comedienne extraordinaire Amy Schumer and former talk-show host Ricki Lake both met their partners via online dating apps. With young people increasingly time-pressured with challenging jobs and arduously long commutes, meeting a partner online seems like the way forward. So what can mathematics tell us about how this world is navigated?

Those who search the stellar skies for little green men (perhaps the most common fictional depiction of extra-terrestrial life) have much more in common with our search for true love than you might think. As we have already discussed in Chapter 5, the Drake equation applies a mathematical framework in an attempt to estimate the number of alien civilisations in our galaxy (1,000 by the lower end estimation of the differing variables).

Although I have been aware of Drake's equation since my childhood obsession with all things space and the cosmos, it was in 2010 that I first saw the quirkiest application of this equation. Step forward Peter Backus, an Economics lecturer at Manchester University but back then a tutor at Warwick University. He published a paper titled 'Why I don't have a girlfriend: An application of the Drake Equation to love in the UK'. Using his redefined parameters of the Drake equation, I will try to calculate the number of women that might be the 'one' for me!

The basic equation is: $G = R \times f_f \times f_{LC} \times f_A \times f_U \times f_P \times L$

Let's go through this item by item:
G = The number of potential girlfriends
R = The rate of formation of people in the UK (in other words, population growth)

I've used data from the Office of National Statistics and a quick bit of Excel manipulation shows the average growth of the UK population since 1960 is 240,000 people.

f_f = Fraction of people in the UK who are female

This is currently about 0.506 in the UK, but I'll be optimistic and round to 0.51.

f_{LC} = Fraction of women in the UK who live in London and Cambridge

I spend a lot of time shuffling between London and Cambridge. The population of London is about 10,657,000 and Cambridge is about 129,000, so this gives a total of 10,786,000. With the UK's estimated population at 66,550,000, this gives 0.16 (10.786/66.550). So that's 16% of the UK population who live in either

London or Cambridge. Realistically, it would be difficult for me to date anyone not living in either of these 2 places due to logistics.

f_A = Fraction of women in my target cities who are age appropriate

Of course, love knows no age boundaries but I am 34 years old, though I probably look a bit younger, especially when I'm clean shaven! So let's say I'm looking for a woman of 25–35 years old, which I estimate to be about 0.2 of the population.

f_U = Fraction of women who are the right age bracket with a university background

I saw this in Peter Backus's original calculation and he acknowledged that there are plenty of intelligent people who didn't go to university. But from my previous experience, I have tended to meet women who have been to university (mainly due to a majority of my circle of friends being graduates). So it's probably a realistic assumption that I'll date someone who has been to university, so a 0.26 according to Backus (though who's to say I won't meet an entrepreneur who left school at 16 to set up their own ventures or someone who fell out of school at 18 for any number of reasons?). From data back in 2012, 27.2% of the population aged 16–74 have a degree or equivalent. So let's use the 0.27.

f_P = Fraction of university-educated and age-appropriate women in my target cities that I am attracted to

Peter Backus had this at 1 in 20 women. He didn't mean to say that the rest of the women aren't attractive, far from it. But he gave an estimate of 1 in 20 that were attractive to him. This is a tough parameter to estimate and can make results swing wildly. As I'm a short guy, taller women perhaps might be more difficult to date! So let's give this a 1 in 10 or rating of 0.1.

L = length of time (years) I have been alive and open to the chance of meeting a potential girlfriend

I am 34, so in theory 34 years. But realistically, I was a very focused school student, and didn't consider dating until after I left school at 18. So this gives us 16 years.

For the rate of formation of people in the UK and my length of time, we can replace this with the more simplistic population of the UK (call it N). The current estimate of the UK population in 2018 is 66,550,000.

$$G = N \times f_f \times f_{LC} \times f_A \times f_U \times f_P$$

Let's now use the values and whack back into the equation, so:

$$G = 66,550,000 \times 0.51 \times 0.16 \times 0.2 \times 0.27 \times 0.1$$
$$G = 29,325$$

The number crunching has given me an estimate of 29,325 women in the UK who meet my initial criteria as potential girlfriend material. That sounds like a reasonable amount, as it's a decent-sized crowd at the old West Ham ground Upton Park, which had a maximum capacity of 35,000. This is 0.044% of the population of the UK, which sounds alright.

Of course, this calculation does not take into account that the woman in question has to find me attractive too and they also have to be single! So this will dramatically reduce the pool of 29,325 women. Let's say 1 in 20 of those 29,325 women find me attractive, half are single and then I get on with only 1 in 10 of them. This reduces the pool as follows: $29{,}325 \times 0.05 \times 0.5 \times 0.1 = 73$

So according to my adapted version of Backus's formulation of the Drake equation, there are 73 women in London (or Cambridge) that I might be able to successfully date. Hmm ... this doesn't sound promising at all. Considering the population of the UK is 66.55 million, the chance that the next person I meet is 1 of my 73 potentials is 1 in 913,000! So if there are 1,000 possible alien civilisations in our galaxy, we are 14 times more likely to find an alien civilisation than I am to find a suitable partner!

A lot of this calculation depends on the ratios you use in the equation, but it is a useful illustration to show the power of maths in helping us to understand something even seemingly unrelated such as the search for love. I am unashamedly an addict of the reality dating show *Love Island*. You'll see me with my official *Love Island* water bottle and reversible 'single'/'coupled up' wristband – currently set to 'single' of course! Mathematically speaking, I am probably in the minority of TV viewers that are in the intersection of a Venn diagram between being both fans of *University Challenge* and *Love Island*. Indeed, if couples from the show end up lonely after the programme, I could come up with a bespoke Drake equation to calculate how many partners they might find in the UK.

What advice can you take from this equation? If you want to increase your chances of finding love, you need to be persistent and keep searching. Spend more time in places where other people who meet your criteria may hang out as well. Opening your minds and relaxing your criteria (for example, my one about being university educated) will open up more possibilities of meeting potential partners.

This suggests that there are 73 women out there for me. (I promise, I haven't tinkered with the numbers, but 73 is a prime number as well, so that number is particularly fun to me!) This brings me to think about the concept of 'the one'. Is there such a thing as 'the one'?

In the UK, our music charts are obsessed with love. Since 1953, there have been more than a hundred chart-topping songs with the word 'love' in their title somewhere. From the 1960s and 1970s we have the Supremes' 'Baby Love' (1964) and Donny Osmond's 'Puppy Love' (1972), and this century, Elton John's 'Are You Ready for love?' (2003) and Ellie Goulding's 'Love Me Like You Do' (2015). Many songs and movies promote the concept of 'the one' – that perfect person out there that is destined for us. Research from the dating site eHarmony shows that 1 in 5 people under 35 believe in the concept of 'the one'.

Randall Munroe, a NASA roboticist and creator of web comic *xkcd*, has actually explored the concept of whether everyone actually only has one soul mate, somewhere in the world. His argument works from the assumption that we are born with a soul mate that is predestined at birth. We know nothing about this person or where they might be. The only thing we know is clichéd, that as soon as we lock eyes with the person, we'll know that they are 'the one'.

However, as our soul mate is just one person ever, we cannot be certain that they are alive. They may have passed away a long time ago or are yet to be born. *Homo sapiens* started walking the Earth around 50,000 years ago. The Population Reference Bureau estimates that around 107 billion people have ever lived. With the current global population around 7.6 billion, this means that 93% of humans ever born are deceased. This means that there is a 93% chance that our soul mate has long since passed onto a heavenly realm. Obviously this excludes humans yet to be born, so this will increase the percentage likelihood that you will not be able to encounter your soul mate in your lifetime.

However, Munroe wants to make the calculations more attainable and says that our soul mate is contemporary and within a few years of our own age. With these parameters, he comes up with an estimate of a pool of half a billion potential matches. But what are the odds of encountering these people?

We often hear people say that you'll know when you meet 'the one', often identifying through eye contact alone. So can maths help us here? The number of strangers we make eye contact with on a daily basis is not easy to estimate. It can depend on your circumstances. If you are in a small town where you know most people, this number can be very small. But if you are a security officer at Heathrow airport, this could be several thousand in a day. And then we would need them to be close to our age. Munroe says that if 10% of the people we lock eyes with are within our age range, then that's around 50,000 people in our lives. With a potential 500 million potential soul mates, the chance of finding true love will only happen in one lifetime out of 10,000. Grim reading again!

The reality is much more nuanced than that. It is of course very fatalistic to believe that we have just 'one' relationship waiting for us out there. The myth of 'the one' is exaggerated by the ideal that a potential partner will fulfil all our needs – social, emotional, physical, intellectual, practical and moral. Those subscribing to 'the one' myth will find themselves on a road to setbacks, misery and disappointment.

At the moment, I'm still relying on chance meetings or introductions by friends in the hope of meeting the right person, but at some stage, I should make a serious effort. How will this serious effort happen? Online dating might be the place to start. The stigma has evaporated.

On the fun, light end of the spectrum of online dating apps, Tinder is a location-based mobile app that allows users to like (swipe right) or dislike (with a flick of the swipe to the left). On the more serious end, websites such as Match and eHarmony are more focused on long-term commitment. In the UK, about a quarter of new relationships are thought to start from an online encounter and £300 million is spent on this industry. It is serious money. Even the mighty Manchester City Football Club, buoyed by their fortune from the Abu Dhabi United Group, have swiped right with Tinder to secure on a multi-year partnership deal.

Assuming they have been well-promoted, online dating sites should have a large number of members so they can work through substantial data about compatible potential partners. Dating websites have to create algorithms that use the data to match me to people they think I might get on with.

All sites will have their own bespoke method for matching those looking for love. However, the American dating site OkCupid have made it very clear that the power of mathematics has been harnessed in their desire to aid your quest for love. They even boast on their site that they 'use math to find you dates' and that they 'do a lot of crazy math stuff to help people connect faster'. OkCupid are so adamant that they are on the right track that they are willing to openly share their algorithm. So what exactly do they do?

The OkCupid website asks potential lovebirds to answer questions and give 3 answers for each particular question. First, users are asked to answer, in a personal capacity, what they think. Second, they state what they would like any potential match to answer. And third, users are asked to allocate a level of importance to that particular question. There are 5 options allowed for assigning the level of importance and the algorithm assigns a numerical value to this.

Level of importance of question	Points value
Irrelevant	0
A little important	1
Somewhat important	10
Very important	50
Mandatory	250

The values attributed to each level impacts the algorithm, but that is what *OkCupid* have chosen. If they changed 'A little important' to 10 points, and 'Very important' to a 100 points, this would impact potential partners being offered.

The Cambridge college where I'm doing my doctorate is Emmanuel College, known affectionately by students and staff (and in wider Cambridge circles) as Emma. So whenever I'm at college, I will tell my family I'm at Emma. My mother is convinced that I'm dating a girl called Emma! So let me perform a compatibility test between myself and a theoretical girl called Emma. (If I do get married one day, a possible venue would be Emmanuel College's beautiful chapel, designed by Sir Christopher Wren. So getting married to Emma at Emma has a ring to it!)

So let's look at my hypothetical *OkCupid* questions:

Bobby	Q1. Do you like long distance running?	Q2. Are you good at cooking?
Self	Yes	No
Wanted from partner	Yes	Yes
Importance	A little	Somewhat
Importance score	1	10

The importance score is $10 + 1 = 11$ points.

Now, let's look at Emma's questions:

Emma	Q1. Do you like long distance running?	Q2. Are you good at cooking?
Self	Yes	Yes
Wanted from partner	Yes	Yes
Importance	Very	Very
Importance score	50	50

The importance score is $50 + 50 = 100$ points.

Based on these we are able to calculate the match between the 2 of us.

My score: as I answered both questions as a 'little important' and 'somewhat important', my questions are worth 11 points. Emma answered as I would have wanted, meaning she receives 11 out of 11 points. So Emma is 100% satisfactory to me. Maybe love is on the cards.

Emma's score: Emma answered 'very important' to both questions, so these questions are worth 100 points to her. I answered yes to only 1 of them, meaning I receive a score of 50 out of 100 points. So I'm 50% satisfactory to Emma. This doesn't sound promising.

This isn't sufficient to calculate compatibility at OkCupid. They will calculate the geometric mean of the 2 percentages. This is a type of mean of average which indicates the central tendency or typical value of a set of numbers, which is useful for data with wide ranges and many data points, like we get on dating website questions. For 2 numbers, we multiply the percentage and find the square root. More generally, we multiply the percentages and find the nth root (n being the number of questions being asked).

Compatibility score = square root of 100×50 = square root of $5{,}000 = 71\%$

This is clearly a simplified version with only 2 questions. In order to give yourself the chance to match with more potential fits, you need to answer as many questions as possible (even if you don't have the patience to sit down and go through 100 of them). The more questions you answer, the more reliable the picture that your data will paint about you. The algorithm seems to only work on questions that are answered by both users.

Interestingly, the website Match.com is not as trusting about its users. It has an algorithm called Synapse, which not only takes into account what your stated requirements are, but also takes into account what you do when on the site. The distance between the 2 is called 'dissonance'. Let's imagine that I tell the website that I'm happy with women who are not football fans but in fact I keep checking out the profile of women who are footy fans. Then the Match.com algorithm will offer up potential women who are football fans too. It looks at what I do, not just what I say I want.

This Synapse algorithm uses the same techniques that are used by internet powerhouses such as Amazon, Netflix and Spotify when they monitor your usage to make recommendations. Feedback is crucial too. The more you rate the suggestions the software makes, the more adept the algorithm is at truly finding matches that might tickle your fancy.

'The course of true love never did run smooth.' These are the words of Lysander to Hermia at the beginning of Shakespeare's *A Midsummer Night's Dream*. There is no such thing as a perfect relationship but mathematics can help me understand how to think about my steps to one day finding that right person in my life. For a start, even with the romantic equivalent of Drake's equation to finding intelligent civilisations in the galaxy, the more I relax my criteria, the higher the potential number of women I might be able to date. Further, the notion of 'the one' is a damaging concept. The more you are open to new experiences and the more you put yourself out there, the higher likelihood of you meeting someone with whom you click.

I personally don't believe that there is one such perfect person, though there are the 'ones for now', people that you might encounter that fit within your circumstances at that time. Then, it's up to you as an individual and a couple to work diligently at keeping the magic alive. Maths can help us understand that it is a numbers game, but you still have to participate with an open mind in the game to find love. For now, if any of my relatives ask my answer is still, 'I will get married soon, but have to focus for now on my doctorate.'

Love Island Cryptic Challenge

On the reality dating show *Love Island* the contestants are set a physical and mental challenge. The first physical challenge involves the letters from the word LOVE ISLAND. The contestants have to use (and contort) their bodies to form the word LOVE ISLAND on the floor of the garden.

Once they have done that, the contestants have their mental challenge. They have to use all the letters from the name of the show again to solve another puzzle, this time a cryptic one.

Can the contestants create a word that represents a fifth version of a warrior king who famously lost in a battle at a place meaning 'hot gates'? As a mathematical clue, this battle took place in the year $2^5 \times 3 \times 5$ BC expressed in prime factor form.

CHAPTER 17

'Sixty Seconds' Worth of Distance Run'

Getting Time to Work in Your Favour

If you can fill the unforgiving minute
With sixty seconds' worth of distance run,
Yours is the Earth and everything that's in it,
And – which is more – you'll be a Man, my son!

These are the closing lines of Rudyard Kipling's 'If'.
(Yes, I know that the poem was written as paternal
advice from Kipling to his son John, but let's extend this
advice to everyone!) Ever since this poem was voted the
UK's favourite in a 1995 BBC opinion poll, the entirety
of this poem has been on display in my bedroom in the
form of a cardboard reinforced poster. In my capacity as
a pastoral form teacher at school, I have also pinned this
poem up on my form class wall, perhaps a subtle form
of positive mindset propaganda for my 11-year-old
tutees.

To me, the poem doesn't just stir up a sense of trying to be the best version of yourself, but also has had a profound impact in terms of my relationship with time itself. The *Oxford Dictionary* defines time as the 'indefinite continued progress of existence and events in the past, present and future regarded as a whole'.

Our best scientific understanding of the universe with the Big Bang theory estimates the age of the universe as 13.8 billion years. In physical cosmology, this measures the length of time elapsed since the Big Bang. Time does keep marching on, one second at a time into the future, not waiting for us. However, I'll accept that at very high speeds approaching the speed of light, things get a little more mixed up. In 'Mr Apple falling on your head' style Newtonian physics, length and time are absolute. In the Einsteinian world of post-general relativity, relativistic time only varies from absolute time when travelling at speeds approaching 299,792,458 metres per second or conditions of extraordinarily high gravity.

Why is time important? It is the one thing given to all of us who have ever existed. Sometimes that period of time can be very brief, but other times fairly lengthy (122 years if you were Jeanne Calment, who passed away in 1997, having seen the painter van Gogh in her father's shop as a child). If you are a biblical figure, then you might reach the ripe old age of 969 as Methuselah apparently did!

When you really think about it, time is the only resource we all have. But that doesn't mean that we need to expunge fun and joy from our lives in our search to maximise our time on Earth. In fact, idle hours in front of the sofa or just chilling in the garden can be an optimal manner to soothe your soul. And if somehow you do manage to devise a method of heading back in time, set the dial to be sipping champagne and nibbling canapes at the late Professor Stephen Hawking's time travellers' party, hosted on 28 June 2009 at Gonville & Caius College in Cambridge (12:00 Universal Time, location: 52° 12' 21' N, 0° 7' 4.7' E).

Scrolling through social media timelines on Facebook, Twitter or Instagram can often just be an exercise in procrastination (well, for me at least). However, on occasion, I do stumble across a post that can be quite startling. One post I chanced upon recently was an image of a 90-year human life (a reasonable life expectancy for anyone with good health) divided up into months. The image had a month represented by a circle, with 36 circles in each row (that is 3 years of 12 months each). There were 30 rows in total, representing the 90 years' life expectancy. This diagram is quite sobering, as you can work out where you are and how much more time you may have (assuming you don't suffer an unfortunate early death).

Here is my life represented by the numbers of months as empty boxes I have left assuming I make it to 90!

CHAPTER 17

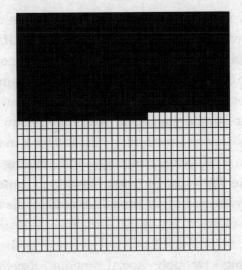

We can often take life for granted, maybe thinking that we have a right to be living and breathing forever, without giving a moment's thought to our ultimate fate. Finding a visual that shows my life split up into monthly circles like that made me take a step back and think about my mortality. Is that a bad thing? Not at all: knowing that life is not a dress rehearsal can give a sense of renewed urgency to do the things you want to do. One of the first times I felt this urgency was during my school days when reading Shakespeare's play *Macbeth*. In what's known in theatrical circles as 'The Scottish Play', the protagonist claims that:

> *Life's but a walking shadow, a poor player, that struts and frets his hour upon the stage, and then is heard no more. It is a tale told by an idiot, full of sound and fury, signifying nothing.*

SIXTY SECONDS' WORTH OF DISTANCE RUN'

I agree that life is brief and that it can pass too quickly, but I vehemently contest his notion that it lacks substance. It is up to us to create meaning and value that enrich our lives, whether through the pursuit of individual goals, family life, career or whatever floats your boat. When I was working as a trader at the investment bank, Lehman Brothers, there was a memorable exchange while getting a lift down to the ground floor on a Thursday evening. Unusually, there was only one other employee in the lift. Elevator lifts are typically a British exercise in avoidance of all eye contact but I shared a smile with this elderly chap. I punctured the silence with, 'Only one more day before the weekend!' His gentle but piercing rebuke was, 'You'll wish your life away if you keep waiting for the weekend.'

I didn't have much time to compute his comment before the lift had reached our destination, apart from a cursory response of, 'I guess you're right.' I did ruminate on his words on my train journey back home. I thought to myself, it wasn't as if I was wishing away the week on a Monday, it was only on Thursday. But he did have a point. If I wished away the 24-hour period from Thursday evening to Friday evening, that would be about 14% of my life vanishing in a flash.

If I were to say to you that every day, you are given £86,400, you would be right in thinking 'that's wonderful' and you would possibly consider waiting for it to accumulate into a substantial pile. But if the condition attached to the daily sum was that it would have be spent every day, then that might change your

241

view on it. The £86,400 represents £1 for every second we have in a 24-hour day. Every day we have 24 hours, and every week 7 days. Rich, poor, hard-working, lazy, lucky, unfortunate – all of us are treated without any prejudice by time; we all have the same 168 hours a week. And each one of us is given 24 hours daily, equivalent to 1,440 minutes or 86,400 seconds. Bearing in mind that most sensible adults will spend about one-third asleep (28,800 seconds), we have only 57,600 seconds to spend each day. So the only question that matters is: how will you spend yours?

Growing up, particularly in my primary school days, the '6–8' routine was embedded as homework time, just as a monk might have reading to do post-vespers (the evening service). At home, after coming back from school and scoffing down some afternoon snacks, the TV would be blazing with *Newsround*, *Blue Peter* and then *Neighbours*. This would finish by 6pm and for the 2 hours before dinner, the expectation was that I and each of my brothers would return to our bedrooms to work on school homework, reading or doing something creative (drawing and art was very much encouraged).

The '6–8' period was subdivided into 4 chunks of 30 minutes each, and in our house there was a concerted push to maximise the quality of time being spent. For us, every 30-minute period was allocated a mark out of 10, like a self-evaluation of our effort. A 10 rating would reflect what I considered to be a missile-guided sense of focus. Conversely, a 0 would represent no work accomplished whatsoever, perhaps because I was completely distracted by something else. The rest was a sliding scale, with 7 being the average aim.

Applying this numerical framework allowed me to really think how effective my use of time was. The use of a quantitative figure enabled the tracking of the ebbs and flows of my working evening. If I felt I had a string of feeble evenings with 3s and 4s, then I would gather my thoughts about what was going awry. Was it motivation? Was it lack of focus? Either way, the ratings per half hour gave the opportunity for regular self-reflection.

The ratings system guided me through my GCSE years at my state school St Bonaventure's and reached its apogee of precision during my 2 years as a Sixth Former on an academic scholarship at the leafy Eton College. There, I divided up the hour into even finer slots of 15-minute portions, thereby increasing the incisiveness of monitoring and rating of my performance. It might seem a bit overkill to the outside world, but this really did help me to maximise how I used my time. When time is a limited resource (it always is), where we have some time dedicated to work and some time to relax, making sure that you are productive when you're working helps you enjoy your free time knowing that you gave it your all. Nowadays, I'm not quite so fastidious about this quantitative self-introspection but in times where I have tight deadlines, I sometimes whip out my old 15-minute slots and the ratings to keep me on track!

It was only in later adult life that I found that this quantitative framework had been written about in some detail. In 2016, Stanford psychologist Cal Newport published *Deep Work: Rules for Focused Success in a Distracted World*. He explains that in the modern world, we are constantly under the bombardment of incoming distractions. Guilty in the dock of distraction are social media in particular, with Twitter, Instagram and Facebook prime assailants.

I do believe that there is absolutely a place for social media in our modern way of life (I'm an active user myself), but it can certainly interfere with your flow. Newport outlines how deep work is the ability to focus, without distraction, on a cognitively demanding task or project. The polar opposite of deep work is shallow work, which is non-cognitively demanding, logistical work that we can do when distracted. An example of this for me is ironing shirts for the week ahead, as it doesn't place significant cognitive demands on me and I can happily listen to the radio at the same time. A frequent favourite of mine is BBC Radio 4's *In Our Time* with host Melvyn Bragg, a weekly radio discussion/podcast that explores the history of ideas. It offers programmes where academics explore ideas ranging from the emancipation of the serfs in 1860s Russia, to the quest to solve Fermat's last theorem in mathematics or even to the pessimistic philosophy of the German Arthur Schopenhauer.

Personally, I have generally managed to make the most of my time, to the point where friends often wonder how I manage to fit so much into my schedule.

There is nothing magical going on apart from the secret that when I work, I actually work. Newport concisely describes the concept of deep work as the super-power of the twenty-first century – the ability to focus single-mindedly on the task at hand. We have had a version of this etched in red ink onto a whiteboard (almost indelibly now) downstairs in our house: 'The ability to concentrate single-mindedly on your most important task, to do it well and to finish it completely is the key to great success.' Newport expressed this in a straightforward quantitative manner:

Work Accomplished = Time spent × Intensity

For example, if you spend 10 hours revising over a day at an average intensity of 3 (so with your phone constantly on, TV on in the background and chatting to friends), you have 30 units of output (3×10). But if you were to ratchet up your intensity to 10 (no distractions whatsoever), you can get the equivalent output completed in 3 hours (10×3). Admittedly, real life is more complex than that, but Newport's big picture still rings true.

My ability to get things done relies on being able to focus. I would be the first to admit social media makes it more challenging than ever for young people to stay on track, but this simple metric of 'intensity' can help you take a step back and check you are really making progress rather than just sitting in front of your work and making little tangible headway. The benefit of focused work is that you can work hard and play hard, so long as you're really productive when you're working!

The hard work component is a key factor in success. With limited time, when you have allocated that resource to work, one must ensure you are working. Angela Duckworth, another American psychologist, has been able to articulate the importance of grafting away at your goals. The University of Pennsylvania professor is renowned for her studies on the concept of 'grit' and self-control. In her 2016 work *Grit: The Power of Passion and Perseverance*, she outlines how grit is a complement to IQ. Grit is about the individual's desire to persevere, show resilience and bounce back from defeat. From researching cadets at an elite American military academy, finalists at the National Spelling Bee and teachers working at tough schools, it was not IQ but grit that led to success. Angela Duckworth argues that the real secret to achievement is not talent (though of course this does exist), but a potent blend of passion and persistence for long-term goals.

Duckworth introduces a formula that can unlock achievements:

Talent × effort = skill
Skill × effort = achievement

Another way of writing this is:

Talent × effort^2 = achievement

What this means is that effort is a square factor more important that talent in achieving things.

In the perennial nature versus nurture debate, I acknowledge that we are born with a spectrum of abilities, which some refer to as talent. But it is our effort that develops this talent into a skill. Your talent may be a natural quickness of feet, but with effort you can develop this into the skill of being adroit at moving a football. Applying effort to this skill results in achievements such as scoring the winning goal in a World Cup final. This framework can aid us when thinking about how best to make the most of our talents. Putting effort in develops that into a skill, and putting effort into the skill gives us achievements. In a world where we seem to genuflect at the altar of genius (sporting, academic, business), it is comforting to realise that we can all improve with this mindset. Duckworth has emphasised the vitalness of effort (it counts twice!) in the goal to achieve your dreams.

As a maths teacher, I genuinely believe that everyone can improve at maths and become more comfortable in dealing with all the numbers, stats and data that life can hurtle our way. However, this does require learners to have a growth mindset. A 'Growth Mindset' is a concept first popularised by Stanford Psychology professor Carol Dweck, which argues that our ability is not fixed. We can rise above what we perceive to be our natural limitations. It has been specifically fine-tuned for mathematics by teacher turned Maths Education professor at Stanford, Jo Boaler. She has been a rock star of the maths education world, and has done sterling work in smashing the myth of being born with a 'maths brain' and the idea that some people innately can or can't do maths. Life is more important than just maths, but changing our attitudes towards this beautiful subject can reflect a more positive can-do attitude towards life.

With only one attempt at this life on Earth, it is up to us to make the most of our talents. In an episode of *The Simpsons*, the father Homer tells his tearaway son Bart, 'No matter how good you are, there's always a million people better.' This is not a problem. We don't have to be the best out there. We just have to be the best version of ourselves. And using our time effectively is critical in doing so.

Mathematics can also help explain the sensation of why we think time seems to pass by quicker and quicker. As a child, waiting for our next birthday or Christmas can seem to take an eternity. In contrast, as adults, we are often astounded at how rapidly these annual landmarks pop up on our doorstop. As a 5-year-old child celebrating a birthday with balloons and cakes, the next year before your sixth birthday takes 20% of your previous life (1 out of 5). As a 10-year-old, waiting for your eleventh birthday consumes 10% of your previous life (1 out of 10). And when you are wise and sensible (hopefully!) at 20, getting to that twenty-first birthday is only 5% of your previous life (1 out of 20). The older you get, the sensation of life coming at you fast is a relative one as it is indeed a smaller percentage of your existence. The older you get, the faster life will fly by. You should not be a bystander as this phenomenon unfolds.

During my English Literature GCSE, we studied one of the most iconic poems from the seventeenth-century metaphysical poet Andrew Marvell, 'To His Coy Mistress'. These lines particularly have resonated with me:

> *But at my back I always hear,*
> *Time's wingèd chariot hurrying near.*
> *Thus, though we cannot make our sun stand still,*
> *yet we will make him run.*

Time will not wait for us as individuals. It keeps ticking on remorselessly. As humans, we want to live an existence as rich in experience as possible. I have found that using a quantitative framework to work effectively lets me live as fully as I can. My life has been a happy and fulfilled one so far, and understanding the relationship between numbers and myself has carved the Bobby Seagull that my friends, family, students and colleagues hopefully value. Mathematics can teach us a lot about understanding our place on a complicated planet, helping to simplify the way in which we can comprehend how the world works. Maths can make our lives better in innumerable ways, though the most important lesson is about making the most of our time here. Ask yourself, what have you done today to make your life and the lives of others around you better than yesterday?

Chapter 17 Puzzle

The Big Bad Wolf and the Three Little Pigs

The Three Little Pigs have built a house together. Obviously, the Big Bad Wolf wants to blow the house down.

From previous experience, the Big Bad Wolf figures that he needs 10 wolves working 15 hours daily for 5 working weeks (only Monday to Friday) to bring the house down.

As the Big Bad Wolf doesn't leave things to chance, he is very meticulous with his time management. If the Big Bad Wolf wants to start this demolition job at 9am and be done in time for tea at 3pm, how many wolves does he need to blow down the house of the Three Little Pigs?

Solutions

Chapter 1 Puzzle Solution

Gareth Southgate's Unusually Weighted Footballs

If the average weight is 80 grams, then we know the total weight is 320 grams as $4 \times 80 = 320$.

If the 3 heaviest weigh 270 grams, then the smallest must be 50 grams (as $320 - 270 = 50$). So that is Jesse Lingard's football.

As the heaviest football is 3 times the weight of the lightest football, Harry Kane's football must be 150 grams (as 50×3 is 150).

The heaviest and smallest add up to 200 grams ($150 + 50$). The 2 middle footballs of Jordan Pickford and Dele Alli are the same weight. So $320 - 200 = 120$. 120 divided by 2 is 60 grams.

So the weights of the footballs are:

Harry Kane = 150 grams
Jordan Pickford = 60 grams
Dele Alli = 60 grams
Jesse Lingard = 50 grams

Now Gareth can get back to coaching the squad!

Chapter 2 Puzzle Solution

Happy Chinese New Year

16 dogs (16 × 4 = 64 feet)
8 roosters (8 × 2 = 16 feet)
4 monkeys (4 × 2 = 8 feet)

64 + 16 + 8 feet = 88 feet in total

Chapter 3 Puzzle Solution

Stamp-collecting Holiday

She's heading for the United Arab Emirates.

If you write down the capital cities of these countries, they are Amman (Jordan), Algiers (Algeria), Addis Ababa (Ethiopia), Accra (Ghana) and Abuja

(Nigeria). These capitals are listed in reverse alphabetical order for states that are are fully independent and sovereign. Abu Dhabi from the UAE would be first in this list.

Chapter 4 Puzzle Solution

Harry Potter and the Quest for 4 Pints of Butterbeer

Thanks to *Die Hard with a Vengeance* for inspiring this puzzle! There are several ways of performing this task and this is only one method.

Let's call the 5-pint tankard Tankard A, and the 3-pint tankard Tankard B.

1. Hermione pours 5 pints in Tankard A.
2. She pours 3 pints from Tankard A into Tankard B. So Tankard A has 2 pints in it.
3. She pours away the 3 pints in Tankard B.
4. Then she pours the 2 pints from Tankard A into Tankard B.
5. She pours 5 pints into Tankard A.
6. Finally she pours 1 pint from Tankard A into Tankard B. So Tankard A has 4 pints in it.

She and her fellow students may now celebrate their OWL results with 4 pints of Butterbeer!

Chapter 5 Puzzle Solution

Astronaut Training for *The Grand Tour* Crew

The capital cities of the countries are:

1. Greece – Athens
2. South Korea – Seoul
3. Iran – Tehran
4. Morocco – Rabat
5. Canada – Ottawa
6. India – New Delhi
7. Paraguay- Asunción
8. Mongolia – Ulaanbaatar

If you write out the first letter of the capital cities, this spells out ASTRONAU. The missing letter is a T to spell the word ASTRONAUT. The capital city in Europe beginning with a T is Tirana, and it is the capital of Albania.

Chapter 6 Puzzle Solution

Damien Hirst's Mathematical Spot Paintings

This is the sequence of dots that Hirst paints: 3, 4, 6, 8, 12, 14, 18, 20.

The clue said that we don't need to be in our prime to solve them. So I was hinting that you should think of your prime numbers which are numbers with 2, and only 2, factors: 2, 3, 5, 7, 11, 13, 17, 19.

If we compare the 2 sequences side by side, it becomes obvious.

3, 4, 6, 8, 12, 14, 18, 20: Hirst's number of dots
2, 3, 5, 7, 11, 13, 17, 19: prime numbers

Hirst's sequence is just the prime numbers with 1 added to them. The next prime number after 19 is 23, so for Damien to continue his sequence, he would paint 24 dots in the next column.

Chapter 7 Puzzle Solution

Ronald McDonald vs Usain Bolt: Chicken McNugget Challenge

By the ninth day, Ronald McDonald will eat more McNuggets.

Ronald McDonald's total after 8 days
$(1 + 2 + 4 + 8 + 16 + 32 + 64 + 128)$ is 255.

Usain Bolt's total after 8 days
$(10 + 20 + 30 + 40 + 50 + 60 + 70 + 80)$ is 360.

On the ninth day, Ronald eats another 256 McNuggets, taking his total up to 511 McNuggets.

Usain eats another 90 McNuggets, taking his total up to 450 McNuggets.

Chapter 8 Puzzle Solution

Eurovision Madness

If you look at the acronym for each of the songs they spell out the following:

> France's song is 'Born a shining star': BASS
> Germany's song is 'People in the crisis hour': PITCH
> Spain's song is 'Time out never ends': TONE
> Italy's song is 'Rock it fallen friends': RIFF
> UK's song is 'New undying love': NUL

The first 4 spell out words to do with music (bass, pitch, tone, riff), whereas the UK's just spells 'nul', quite literally nul points!

Chapter 9 Puzzle Solution

Joe Wicks's Mathematical Workout Routine

We assign the letter A = 1, B = 2, C = 3 all the way through to X = 24, Y = 25, Z = 26. Then we assign number values to the first letter of each routine as follows:

Burpee = 2
Calf Raise = 3
Lunge = 12
Mountain Climber = 13
Press Up = 16
Squat = 19

We can see that the number represents the amount of times that particular routine should be performed. So as 'tuck jump' begins with the letter T, this equates to a 20. So I should perform 20 tuck jumps. (Note that 20 cleanly performed tuck jumps at the end of all the above is quite demanding!)

Chapter 10 Puzzle Solution

James Corden and Russell Brand on a Big Day Out at West Ham

This is a classic question about reverse percentages that many GCSE students will be familiar with. If the discount gives you 20% off, this means that the current price is 80% of the original.

Therefore £64 = 80%. We can work out that £0.80 = 1% and thus £80 = 100%.

So the pre-sale price of the 2 items in total is £80. As the shirt costs £40 more than the scarf, you either set up a simple equation or solve through trial and error to work out that pre-sale price of the shirt is £60 and the scarf £20.

For those interested, the simple equation could be as follows. Let X = the price of the scarf. Then 3X = the price of the shirt.

$$X + 3X = 80$$
$$4X = 80$$
$$X = 20$$

So the pre-sale scarf was £20 and the shirt is 3 times as much, which is £60.

Chapter 11 Puzzle Solution

Marty the Magician's Secret Stock Market Method

If the share price doubles every day and you are investing from day 7, you will benefit from 3 days of doubling.

At the start you have £10,000. Here is your total amount at the end of each day:

Day 1: £20,000
Day 2: £40,000
Day 3: £80,000

Or you could use indices. If a share price doubles, this means a 100% increase in the share value each day. So if the original share price is 100%, then a 100% increase leads the share price to 200%, a factor of 2.

As it is for 3 days, you can do $2^3 = 2 \times 2 \times 2 = 8$. So $8 \times £10,000 = £80,000$ for the final value of your shares. Indices would be much more efficient than manually calculating if you had many more days in your calculation.

So the stock is worth £80,000 but as your initial investment was £10,000, you have made a £70,000 profit.

Chapter 12 Puzzle Solution

Catch Me If You Con : The Game Show with a Panel of 10 Star Hosts

Task 1:

Jeremy Paxman – *University Challenge*
Jeremy Vine – *Eggheads*
Sandi Toksvig – *QI*
Bradley Walsh and Shaun Wallace – *The Chase*
Richard Ayoade – *The Crystal Maze*
Nick Hewer and Rachel Riley – *Countdown*
Richard Osman and Alexander Armstrong – *Pointless*

Task 2:

For the first position, there 10 different possibilities. Then for the second, there are 9. Then for the third, there are 8. We keep multiplying $10 \times 9 \times 8$ and so on.

This is mathematically called factorial, so we calculate:
$10! = 10 \times 9 \times 8 \times 7 \times 6 \times 5 \times 4 \times 3 \times 2 \times 1 = 3,628,800$

So there are 3,628,800 different ways in which the 10 stars could line up to ask their questions.

Chapter 13 Solution

Emmanuel College Porters and Probability

There are a couple of ways to solve this.

First is direct probability. You can multiply out the chance of picking the correct letters in order as: $2/8 \times 2/7 \times 1/6 \times 1/5 \times 1/4 \times 1/3 \times 1/2 \times 1/1 = 4/40,320 = 1/10,080$

There are 2 chances out of 8 to pick the first E correctly. Then there are 2 chances out of 7 to pick the first M correctly, then 1 chance out of 6 to pick the second M correctly and so on.

The other way involves using basic combinatorics. EMMANUEL is an 8-letter word. If there were no repeating letters, the number of arrangements would be: $8! = 8 \times 7 \times 6 \times 5 \times 4 \times 3 \times 2 \times 1 = 40,320$

However, since there are repeating letters, we have to divide to remove the duplicates accordingly. There are 2 Es and 2 Ms.

So $40,320/(2! \times 2!) = 40,320/4 = 10,080$ ways of arranging it. So the probability is $1/10,080$.

Chapter 14 Puzzle Solution

Post-match *University Challenge* Handshakes

There are 36 handshakes in total.

For 9 people in total, the first person shakes hands with everyone else. That will total 8 handshakes (since they do not shake hands with themselves).

The second person now shakes hands with everyone, except herself and the first person, with whom she has already shaken hands. This will add 7 handshakes to the total.

Continuing along the line, each person will then add 6, then 5 and so on until the ninth person, who will already have shaken hands with everyone else, and will add 0 handshakes to the total.

The total handshakes is therefore: $8 + 7 + 6 + 5 + 4 + 3 + 2 + 1 = 36$

Chapter 15 Puzzle Solution

Friends Party at Tate Modern

During the party I play the following:

$3 \times$ Ellie Goulding
$1 \times$ Big Shaq
$4 \times$ Kerry Andrew

SOLUTIONS

1 × Childish Gambino
5 × Stormzy
9 × Jamie Cullum
2 × Muse

If you write out the numbers in order you get 3141592. To mathematicians, this string of numbers should look familiar. If you insert a decimal point after the first number you get 3.141592 and this is recognised as the first 7 digits of π (the mathematical constant which is the ratio of a circle's circumference to its diameter). The next digit of π is 6 and so I need to play 6 of Sigrid's pop songs to conclude this most excellent party!

Chapter 16 Puzzle Solution

Love Island Cryptic Challenge

$2^5 \times 3 \times 5$ BC expressed in prime factor form is 480 BC ($2 \times 2 \times 2 \times 2 \times 2 \times 3 \times 5$).

In 480 BC the famous battle of Thermopylae (which means 'hot gates') took place in Greece. The film *300* saw the Persian King Xerxes (played by Rodrigo Santoro) defeat 300 Greek warriors, nobly led by Gerard Butler's character, the legendary King Leonidas.

As the clue said it represents the fifth version, we can use the Roman numeral of V for 5.

SOLUTIONS

You can rearrange the letters from the words LOVE ISLAND into LEONIDAS V. Good luck to the contestants in being able to physically rearrange themselves on the garden floor to form this word!

Chapter 17 Puzzle Solution

The Big Bad Wolf and the Three Little Pigs

How many wolf hours are needed?

Each wolf will work 15 hours × 5 weeks × 5 days = 375 wolf hours.

As there are 10 wolves, it will be 10 × 375 = 3,750 wolf hours in total.

If he wants to demolish from 9am to 3pm, this is 6 hours.

3,750 wolf hours ÷ 6 hours = 625 wolves

(This is quite handy as 625 is 25 × 25, so the Big Bad Wolf can assemble all of them in a square of 25 rows and columns!)

Acknowledgements

There are so many people I would like to thank for supporting me on my journey in general, so I apologise if I have left someone out as I am writing this at 2am!

First and foremost, my family who have been with me since the very beginning and encouraged me to pursue my dreams. My cousins, uncles, aunties and my brother's extended family for plenty of fun times.

My agent Robert Gwyn-Palmer who was the first person to suggest that I should write a book. I wouldn't be here without you and thank you for continued support on our journey.

My editor, Jamie Joseph, who has shown saintly patience in dealing with me and his superb editing skills, my copy editor Paul Simpson who has helped me cross the finish line and the rest of the team at Penguin Random House including Chloe Rose and Caroline Butler. Thanks also to Jack Beattie, the engineer with whom I recorded the audio book.

ACKNOWLEDGEMENTS

My accountants Nigel Haigh and Mark Whetren at Beavis Morgan, for making sure I'm on top of my finances.

People and institutions that have supported my education. St Michael's (Mrs Lang, Ms Henderson), Nelson, Hartley, St Bonaventure's (Sir Michael Wilshaw, Nick Christie, John Workmaster, Matt Farley, Di & Paul Halliwell), Eton College (Peter Broad, Kit Anderson, John Lewis, Tony Little, Gary Savage, Joe Spence, Anthony Dean, Bob Hutton), Lady Margaret Hall (Gabrielle Stoy, Alan Rusbridger), Royal Holloway (Christine Farmer, Michael Spagat), Hughes Hall (Carole Sargent), Emmanuel College (Dame Fiona Reynolds, Jeremy Caddick, David Livesey and all the amazing porters – Dave, Paul, David, Monty, John, Irene, Daniel).

All my students who I have ever taught (or tried to teach!) at Chesterton Community College, the East London Science School and Little Ilford School (and the head teacher Ian Wilson).

The head of my Cambridge Doctorate of Education course Elaine Wilson and my direct supervisors Steve Watson and Julie Alderton. Also to Mark Dawes, my co-leader for my PGCE course, and my PGCE mentors Nayeerah Aullybux and Matt Woodfine. To Shirley Conran for inspiring me and sharing her work on maths anxiety with me.

To my best ever friend, Jennifer Dyson-Batliwalla, and her husband Ben, for all their continued love and support. To all my other friends including Alex Holmes, Alice Rees, Alvin Ross-Carpio, Andrew Perera, Andrew Philip, Anne Catherine-Sternberg, Anneka Brown, Anureet

ACKNOWLEDGEMENTS

Johal, Asma & Mo Ali, Atlanta Plowden, Barri Mathuru, Ben & Rachel Maddison, Beth & Michael McGregor, Bob Chapman, Charlotte de Brabandt, Chris & Jamie Old, Chris Stainton, Christiane Kienl, David Baynard, David & Becky Garbutt, David Leiser, Donna Golach, Emily Dinsmore, Emma Neale, Farah Mahammoud, Gareth Sturdy, George Robinson, Georgie McHale, Ginny Rutten, Gosia Stanislawek, Grace Le, Hannah Walker, Haydar Koher, James Daly, James Gray, Jay & Kiran Keshur, Jason Ash Deacon, Jeremy Judge, John Kirk, Karen Garner, Kate Stockings, Kate Davies & Phil Martin, Katie Gaynor, Kay Blaney, Keon & Heidi Manouchehri, Krystyna Zawal, Laurent de Brabandt, Liang Long-Ho, Matt & Laura Flitton, Matt Baron, Matt Dodd, Matt & Gulnoor Jones, Matt Marshall, Matthew Osment, Melissa Varney, Nalini Cundapen, Natasha & Eddy Hall, Neha Chaudhary, One Seok-Lee, Pamela R, Philippe & Eimear Hoett, Rachel Evans, Rachel Johnson, Rafi Anwar, Richard Hume, Richard Freeland, Shermaine Sea, Sophie Warner, Stefanie Daley, Susanna Kuruvilla, Tom Corby, Tom Grady, Vickesh Sharma, Will Southgate, Wolfram Bosch.

My sporting supporters at East End Road Runners, the Cambridge University Hare & Hounds, West Ham football club (including Rob Pritchard) and the team at YouTube's Football Daily channel as well as my East Ham fitness instructors Dave McQueen and Heather Thomas.

My friends at Label1 (Lorraine Charker Phillips, Simon Dickson, Jo Taylor, Merle Currie, Sam Palmer, Kat Bannon, Vusi Kaolo, Sean Valentine, Adam Scott,

Ben Ring, Malcolm Remedios and others) and the team at Talkback.

My oldest students ever, my BBC GCSE Maths Challenge team Naga Munchetty, Jayne McCubbin, Tim Muffett and producer Bella McShane.

The team at National Numeracy including Mike Ellicock, Abigail Hartfree, Lauren Street and Rachel Riley.

Radio 4's Lauren Harvey for allowing me to contribute regularly to *Puzzle for Today* and my producer Dave Edmonds on our *Polymaths* show.

My friend and editor at *FT Money*, Claer Barrett.

My friends at Newham Money Works and the borough of Newham London more widely, and those who I worked with including councillor Veronica Oakeshott and locals Cecilia & Gerard Walsh.

Organisations that I have worked closely with such as OxFizz, UpRising, the Open University, the Bright Ideas Challenge team at Shell (Marcus Alexander-Neil and Hannah Furrer) and the Itza team (including Anthony Bouchier).

All my previous employers, such as the Muirhouse Youth Development Group, KPMG, Lehman Brothers, Nomura, PwC.

Maths writers that have particularly inspired me including Alex Bellos, Eugenia Cheng, Hannah Fry, Jo Boaler, Marcus du Sautoy, Rob Eastaway and Simon Singh.

Those who supported me on my first book, *The Monkman and Seagull Quiz Book*, including the Eyewear

ACKNOWLEDGEMENTS

publishing team (Todd Swift), Janina Ramirez, Kevin Ashman, Louis Theroux and Stephen Fry.

All my friends who have got me into quizzing at Cambridge and the Quiz League of London. My *University Challenge* team of Bruno Barton-Singer, Tom Hill and Leah Ward. The Emmanuel quiz legends Alex Guttenplan and Jenny Harris. The Emma University Challenge teams that have followed including Alex Mistlin, Ben Harris, Connor MacDonald, Dani Cugini, Edmund Derby, James Fraser, Kitty Chevallier, Sam Knott and Vedanth Nair. My Wolfson pal Eric Monkman and his teammates Ben Chaudhri, Paul Cosgrove, Justin Yang and Louis Ashworth. My friends at Peterhouse who inspired me to quiz, Hannah Woods, Oscar Powell, Julian Sutcliffe, Thomas Langley. Tomas Kesek and Evan Lynch for starting me on my journey and other quizzers including Ellie Warner, Oliver Sweetenham, Ephraim Levinson, Natasha Voakes, Xiao Lin, Afham Raoof, Josh Pugh-Ginn, Boin Cheng, Daoud Jackson, Matt Nixon, Ian Bayley, Oliver Garner, Dougie Morton, Paddy Duffy, Martyn Smith. And of course the *University Challenge* staff including Jeremy Paxman, Roger Tilling, Tom Benson, Lynne, Toby Hadoke and others.

Based on probability, it is nearly 100% certain I have missed some important names and I can only apologise. I'll make sure I mention you in my next book!

References

Over the years, my mathematical thinking has been inspired and influenced by many popular maths books. Apart from a multitude of general online resources, these are some of the writers and books that I owe a debt of gratitude. Apologies if I've missed any authors or books out here!

Acheson, David, *1089 and All That: A Journey into Mathematics* (2010)

Bellos, Alex, *Alex's Adventures in Numberland* (2011)

Benjamin, Arthur, *The Magic of Maths : Solving for x and Figuring Out Why* (2015)

Beveridge, Colin, *Cracking Mathematics: You, This Book and 4,000 Years of Theories* (2016)

Boaler, Jo, *The Elephant in the Classroom: Helping Children Learn and Love Maths* (2015)

Boaler, Jo, *Mathematical Mindsets: Unleashing Students' Potential through Creative Math, Inspiring Messages and Innovative Teaching* (2016)

REFERENCES

Butterworth, Brian, *The Mathematical Brain* (1999)

Cheng, Eugenia, *How to Bake Pi: Easy Recipes for Understanding Complex Maths* (2016)

Christian, Brian and Griffiths, Tom, *Algorithms to Live By: The Computer Science of Human Decisions* (2017)

Cowley, Stewart and Lyward, Joe, *Man vs Big Data: Everyday Data Explained* (2017)

Devlin, Keith, *Life by the Numbers* (1999)

du Sautoy, Marcus, *The Number Mysteries: A Mathematical Odyssey through Everyday Life* (2011)

Duckworth, Angela, *Grit: Why Passion and Resilience are the Secrets to Success* (2017)

Eastaway, Rob and Haigh, John, *The Hidden Mathematics of Sport* (2011)

Eastaway, Rob and Wyndham, Jeremy, *How Long Is a Piece of String?* (2003)

Eastaway, Rob and Wyndham, Jeremy, *Why Do Buses Come in Threes?: The Hidden Maths of Everyday Life* (2005)

Ellenberg, Jordan, *How Not to Be Wrong: The Hidden Maths of Everyday Life* (2015)

Fry, Hannah, *The Mathematics of Love (TED)* (2015)

Gould, Stephen Jay, *Ever Since Darwin: Reflections in Natural History* (1977)

Gullberg, Jan and Hilton, Peter, *Mathematics: From the Birth of Numbers* (1997)

Levitin, Daniel, *A Field Guide to Lies and Statistics: A Neuroscientist on How to Make Sense of a Complex World* (2018)

REFERENCES

Mandelbrot, Benoit B. and Hudson, Richard L., *The (Mis) Behaviour of Markets: A Fractal View of Risk, Ruin and Reward* (2008)

Matthews, Robert, *Chancing It: The Laws of Chance and How They Can Work for You* (2016)

Newport, Cal, *Deep Work: Rules for Focused Success in a Distracted World* (2016)

Oakley, Barbara, *A Mind for Numbers: How to Excel at Math and Science (Even If You Flunked Algebra)* (2016)

Revell, Timothy, *Man vs Maths: Everyday Mathematics Explained* (2017)

Rooney, Anne, *The Story of Mathematics* (2009)

Singh, Simon, *The Simpsons and Their Mathematical Secrets* (2014)

Stewart, Professor Ian, *Professor Stewart's Incredible Numbers* (2016)

Stewart, Professor Ian and Davey, John, *Seventeen Equations that Changed the World* (2013)

Strogatz, Steven, *The Joy of X: A Guided Tour of Mathematics, from One to Infinity* (2014)

About the Author

BOBBY SEAGULL became a cult star as a contestant on *University Challenge* in 2017. He is now an obsessive quiz enthusiast, TV presenter, secondary school maths teacher, a doctoral student researching maths anxiety at Cambridge University and also a long-suffering West Ham fan.

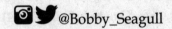 @Bobby_Seagull